Virginia Civil War Battles and Leaders Series

Lexington and Rockbridge County in the Civil War

2nd Edition

Robert J. Driver, Jr.

HERITAGE BOOKS
2023

HERITAGE BOOKS
AN IMPRINT OF HERITAGE BOOKS, INC.

Books, CDs, and more—Worldwide

For our listing of thousands of titles see our website
at
www.HeritageBooks.com

Published 2023 by
HERITAGE BOOKS, INC.
Publishing Division
5810 Ruatan Street
Berwyn Heights, MD 20740

Heritage Books by Robert J. Driver, Jr.:
Augusta County, Virginia Confederate Soldiers
Augusta County, Virginia Confederate Soldiers: Photo Pages
Confederate Sailors, Marines and Signalmen from Virginia and Maryland
Confederates of Prince Edward County and Deaths in Farmville Hospital, 1862–1865
First and Second Maryland Cavalry, C. S. A.
First South Carolina Cavalry
The First and Second Maryland Infantry, C.S.A.
The Virginia Regimental Histories Series: 14th Virginia Cavalry, 2nd edition
Virginia Civil War Battles and Leaders Series: Lexington and Rockbridge County in the Civil War, 2nd Edition

International Standard Book Number
Paperbound: 978-0-7884-3028-2

Lexington And Rockbridge County In The Civil War

This series is dedicated to the State of Virginia and all of her people who lived during the War Between the States. It is the purpose of this series to preserve, as a part of our heritage, the deeds and sacrifices of these Virginians. Your support of this project is greatly appreciated.

CONTENTS

Acknowledgements

The history of Lexington and Rockbridge County during this period could not have been written without the assistance of the people of the community.

The Clerk of Court of Rockbridge County and his staff gave me access to the county records in their possession.

The Lexington City Manager and his staff made available the Minute Books of the Lexington Town Council for the period.

Royster Lyle of Lexington made me aware of the John Letcher Papers in the Marshall Museum Library (now in Rockbridge Historical Society papers) and shared his collection of material on "Hunter's Raid."

The librarians of the Virginia Military Institute, Washington and Lee University and the Rockbridge Regional Library were extremely helpful in locating rare publications, documents and other materials.

My special thanks to Lisa McCown of the Washington and Lee University Library's Rare Book Room, who doubles as Custodian of the Rockbridge Historical Society Collections, for locating many documents and photographs in the two repositories.

Another special thank you to Diane Jacob of the V.M.I. Archives for locating the records of that institution during the War Between the States.

Harold Howard of Lynchburg deserves a special kudo for his confidence and support in this endeavor.

Lowell Reidenbaugh has earned a special thank you for editing the manuscript.

Richard Armstrong took time out from his writing to do the maps for this endeavor.

As always, my wife Edna has earned my heart-felt thank you for her proofreading and other assistance in this project.

FORWARD

The writing of the history of Lexington and Rockbridge County during the War Between the States has been a true labor of love. My effort has been to give a fair and impartial account of the causes and effects of the Civil War, as it related to the people at home during the bloody four years it took place. Material on some aspects of the home front during this period was not available. The history of the individual soldiers in the different companies from Rockbridge is being recorded in other books of the Virginia Regimental Series. This is not a history of Washington College or the Virginia Military Institute during the war. Because of their close connection with the community, and the wealth of published material concerning their participation in the conflict, and the scarcity of reports on local Reserve and Home Guard units, and other activities, the reader may perceive it that way. Every effort has been made to use first hand reports on actions and events, rather than my personal feelings and beliefs on the subject.

CHAPTER I

"Like the muttering thunder before the storm"

John Brown's raid on Harpers Ferry in October 1859 sounded an instant alarm in the peaceful Shenandoah Valley town of Lexington and surrounding areas of Rockbridge County. Brown's abortive attempt to seize the federal arsenal and arm slaves for an insurrection stirred a flurry of activity among the local units of the state militia. Excitement subsided, however, with the speedy trial and execution of Brown at Charlestown, which was attended by a detachment of Cadets from the Virginia Military Institute in Lexington, the only military force from the area to attend the hanging.[1]

Of far greater significance to the residents of Rockbridge than the John Brown incident was the widening rift between political factions in Virginia and other sectors of the United States. John Letcher, a Lexington lawyer and pro-Unionist, narrowly won the governorship of the state over William L. Goggin of Bedford County. A vigorous advocate of states' rights during his years in the United States Congress, Letcher attempted to skirt the secession issue during the hotly contested gubernatorial race. With only minimal support from eastern Virginia, "Honest John" released a pre-election statement in which he promised to resist any movement of federal troops across Virginia to enforce "unjust, iniquitous and unconstitutional laws, either in Virginia or any other state." His margin of victory was only 5,569 votes. Letcher, who failed to carry Rockbridge County, received most of his support from the rest of the Shenandoah Valley and western Virginia.[2]

The national election of 1860 was a bitterly fought contest in which Abraham Lincoln of Illinois, championed by the abolitionist elements, was elected president over three conservative candidates, although receiving nearly one million fewer popular votes than his opposition. By splitting the conservative vote three ways, Lincoln was able to capture the presidency in the electoral vote. The Rockbridge electorate cast most of its votes for John Bell of Tennessee, who ran on the platform of "The Constitution, the Union and the enforcement of the laws." John C. Breckinridge of Kentucky ran second and Stephen A. Douglas of Illinois a distant third at the county polls. Lincoln received one vote in Rockbridge County.[3]

To Southerners, Lincoln's election was tantamount to an utter disregard for the Constitution, the judgments of the Supreme Court and, consequently, the vested rights of the Southern people. While most Virginians remained loyal to the Union, the enthronement of Lincoln aroused secession sentiments among many in the Old Dominion, just as it did in its sister states below the Mason and Dixon line. The movement in Virginia was led by ex-Governor Henry A. Wise, chief protagonist of a small but vocal minority. Despite Letcher's efforts to calm the unrest, the demand for a special session of the General Assembly to debate the issue was too strident to ignore and the extra session was convened on January 7, 1861.[4]

While the citizens of Rockbridge manifested a general calm during this period, interest in military affairs escalated. Drills took on fresh meaning and there was a noticeable improvement in attendance at these events.

Meanwhile, other signs of the approaching storm were beginning to appear. The Lexington town council voted to pay $25 to rent a drill room for the "Rockbridge Rifles." The Baltimore firm of Canfield & Brothers advertised military goods in local newspapers. The Franklin Debating Society, organized in 1810, was summoned to meetings by the ringing of a bell in the streets and argued at length the issues of the day, including secession. At Washington College, a student from the newly seceded state of South Carolina returned from Christmas vacation wearing a badge of secession that had been introduced in the Palmetto State. It was a blue cockade that quickly found favor among Lexington residents.

There also were non-military matters to occupy the attention of Lexington residents. John C. Middleton was reelected mayor. His trustees for the new term were William C. White, William McLaughlin and Colonel John M. Ruff.[5]

When the Virginia legislature met in January 1861, it authorized the election of delegates to a state convention, a system employed by other southern states to consider secession. Letcher, who had hoped for a political compromise at the forthcoming peace convention in Washington, reluctantly issued the necessary proclamation. The peace convention was foredoomed to failure, however, because of the absence of representatives from seceded states.[6]

As states in the Deep South ratified the ordinance of secession, more and more young people of Rockbridge donned secessionist badges. The great majority of these were students at Washington College and V.M.I. who, with their lady friends, sang lustily and with feeling the popular airs of the time such as "Dixie" and "The Bonnie Blue Flag."

The election of county delegates to the state convention was debated vigorously. On February 4, the same day that the Confederate States of America took form in Montgomery, Alabama, Rockbridge voters elected as their delegates Samuel McDowell Moore and James B. Dorman, both strong Unionists, over John W. Brockenbrough, who leaned toward secession, and Cornelius C. Baldwin, an avowed secessionist.[7]

Delegates to the convention gathered in Richmond on February 13. While sentiment generally favored Virginia remaining in the Union, other forces were working in the opposite direction. The situation at Fort Sumter in Charleston Harbor was worsening almost daily. Lincoln's inaugural address on March 4 did nothing to allay fears of Virginians. Lobbyists from seceded states beseeched Letcher to seize the arsenal at Harpers Ferry as well as Fortress Monroe and the Gosport Navy Yard at Norfolk. To all the entreaties, Letcher turned a deaf ear. He steadfastly refused to be coerced into action against the United States government.[8]

In Lexington, some thirty students at Washington College petitioned the administration for a military class to be drilled by a Cadet from the V.M.I., if the faculty would assume responsibility for the arms. The petition was rejected. The faculty, it was announced, did not wish to be responsible for the arms and "...it will not prove useful, and will bring a rabble crowd to the College premises which the Faculty and students would wish to avoid." A week later the faculty reconsidered its decision. It voted approval provided no firearms were brought on campus, there was no drilling during study hours, the commander of the company met their approval and that the faculty might prohibit drill if "... it interferes with the proper discharge of college duties."

Alexander T. Barclay, a student at the college, reported the memorable events:

> Those piping days of peace, those glorious spring days of 1861. Nature never looked more lovely, the blue mountains and green fields of old Rockbridge never more beautiful; all nature seemed at peace. But soon the ominous portents of the times gave forecast of the storm that was to break with hellish fury upon our devoted Southland. The public mind was being fired for war. The daily press fed the flames. I recall with what anxiety we awaited the arrival of the stage from Goshen — we had no railroads or telegraph lines to Lexington. We sat up till midnight for that tardy coach and spent the next hour or so reading the debates in Congress. Davis, Benjamin, Hill, Toombs, Stephens, Breckinridge and others were championing the cause of the South. Soon events began to march so rapidly that we scarce had time to comment before something more absolute and direct attracted attention.
>
> The spirit of the times now began to show itself in our local history. Union and secession became pregnant, living words. Flags were raised, speeches were made, companies were organized and drilling began. The first symptom to manifest itself among the boys of Washington College was in

declamation, which we had every Friday afternoon. The president of the college was a pronounced Union man, and he discountenaced all declamation which he was pleased to call treasonable, with a large T. The boys attempted to introduce speeches of the orators in Washington, but were ordered 'Sit down, sir.'[9]

As war fever mounted in Lexington, the debate over secession waxed hotter at the convention in Richmond. On April 1, delegate Dorman of Rockbridge introduced a resolution submitting the secession question to a vote, which was conducted on April 4. The measure was defeated convincingly by a margin of 88 to 43. Later a delegation was sent to Washington to confer with President Lincoln, who assured his visitors, among other things, that he would employ force against the seceded states if they persisted in their errant ways.[10]

The firing on Fort Sumter by South Carolina troops on April 12 supplied fresh fuel to the secession clamor in Virginia. When a militant throng gathered outside the governor's mansion in Richmond, it was reminded by "Honest John" Letcher that the Old Dominion was still in the Union. He calmly urged the demonstrators to disperse. As they did so, the Confederate flag that had been raised over the mansion was taken down.[11]

When news of the Fort Sumter bombardment reached Lexington, secessionists erected a flag pole on the courthouse lawn. Soon the Confederate banner fluttered from the top of the staff, a bonfire was kindled and speeches were delivered. The *Valley Star* of Lexington reported:

> On Saturday last at eleven o'clock, the flag of the Southern Confederacy, bearing fifteen stars and the motto, "Union of the South," was run up in front of the Court-House amid the cheering of the assembled multitudes. Speeches were made by J.[ames] G. Paxton, J.[ames] W. Massie, Major [Raleigh E.] Colston, J.[John] B. Brockenbrough, and J.[ames] C.[ole] Davis.
>
> In the afternoon of the same day our Union friends attempted to erect a monster pole, but it broke into pieces of its own weight. This disaster, together with the news received almost at the same moment that war had been inaugurated at the South, induced the secessionists to down their flag on Monday morning with the view of removing every difficulty in the way of uniting all parties in the great crisis. May God avert the calamity of war, — but if it must come, may God unite us in defense of our homes!
>
> A serious difficulty came near occuring on Saturday afternoon. A fight took place, and the report spread along the crowd-

ed street that the secessionists and Unionists were about to have a battle. All rushed to the spot, — News reached the Institute that some cadets were in danger of being killed. The corps turned out in a body under arms and came marching up in battle array. On the other hand the volunteers were called out and the citizens were very generally preparing to defend themselves, — when, happily, milder counsels prevailed and all parties retired quietly to their homes. May civil strife always end thus.[12]

Immediately after the incident Cadet Alexander C. S. Gatewood reported:

> Some of the secession *Citizens* invited the Cadets to come up and help raise a secession *flag* and of course [we] excepted (sic) the invitation and went up on Saturday morning and raised the flag. All went well and had several good speeches from some of the most worthy citizens. (In the meantime some of the rowdies[,] ruffians[,] town Mechanics) were raising a *Union Flag* when they got it very near up it broke [into] three or four pieces. That made them mad and they wanted to *pick* a *quarrel* with some of the Cadets which they did after insulting some of them very *cowardly.* They drew some pistols on the Cadets and the boys pitched into them [and] come very near whipping them but [there] was such a crowd of these rowdies that they were too many for the Cadets. One of the Cadets. . . ran down to the *barracks* [and] beat the drum and hollowed "Cadets Turn Out" it was ten minutes until every one of us had on our accoutrements our guns loaded and were down ready to fight. We all started off to town. Col. Smith saw [us] he ran accross the field and met us. Besides himself was Major [William] Gilham, [Raleigh] Colston and every Professor down there. You cannot imagine the excitement. . . . Col. Smith and all the Profs were then [there?] it kept them as busy as they could be to get Cadets to come back to barracks. You cannot imagine how it was. Here they came two hundred and twenty Cadets with their guns loaded. The only thing that kept us back was that we thought we were up in the midst of town we might kill some innocent persons. The women were crying for their husbands and children. Col. Smith had employed Mr. Miche and [James W.] Massie to investigate the matter. He says we shall have our rights if we cant get them here. He will have it carried to the *Legislature.* I tell you "Spix" [Colonel Smith] is all right. He has made two or three speeches to us here in a few day's. He says if Va dont go out he is going out. I hope you are a Secessionists. The *13th day of April* is *a day long to be* "Remembered."[13]

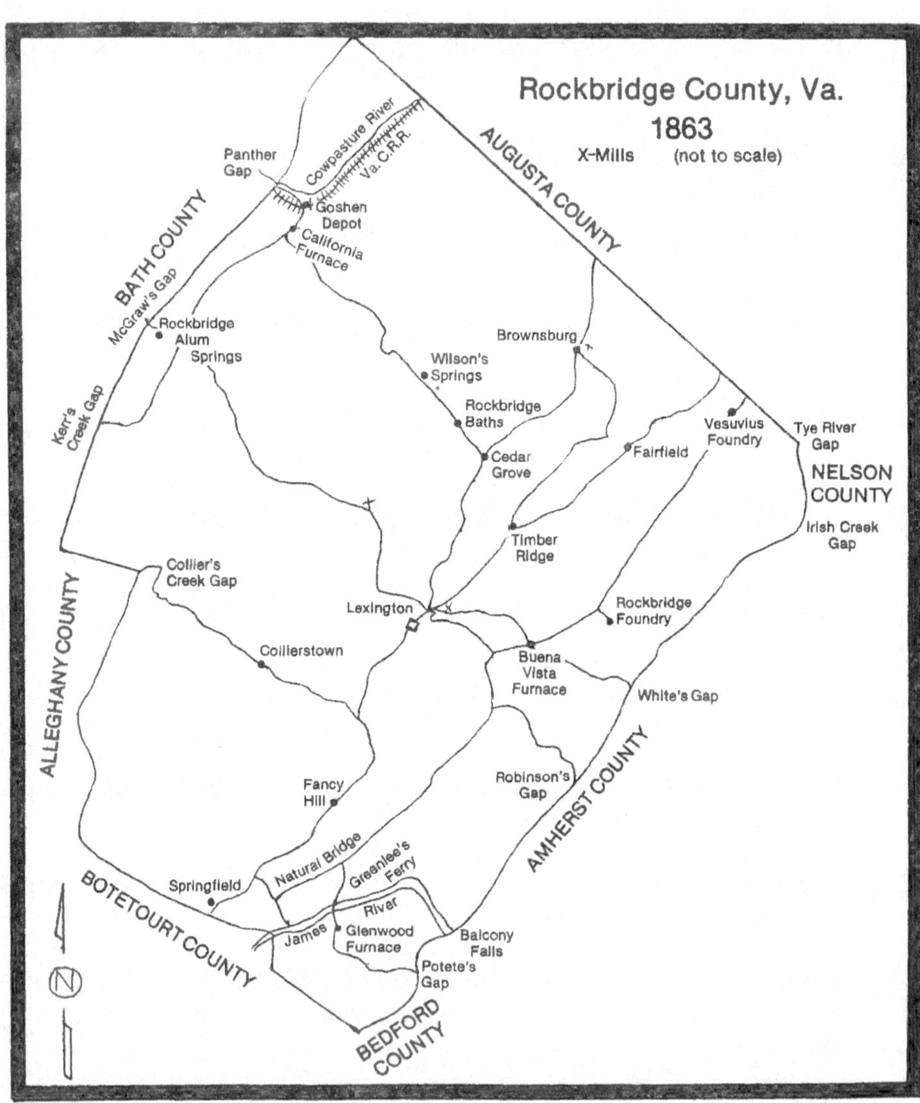
Rockbridge County, Va.
1863
X-Mills (not to scale)
AUGUSTA COUNTY
BATH COUNTY
ALLEGHANY COUNTY
BOTETOURT COUNTY
BEDFORD COUNTY
AMHERST COUNTY
NELSON COUNTY
Panther Gap
Cowpasture River
Va. C.R.R.
Goshen Depot
California Furnace
McGraw's Gap
Rockbridge Alum Springs
Kerr's Creek Gap
Brownsburg
Wilson's Springs
Rockbridge Baths
Cedar Grove
Vesuvius Foundry
Tye River Gap
Fairfield
Irish Creek Gap
Timber Ridge
Collier's Creek Gap
Lexington
Rockbridge Foundry
Collierstown
Buena Vista Furnace
White's Gap
Robinson's Gap
Fancy Hill
Natural Bridge
Greenlee's Ferry
Springfield
James River
Glenwood Furnace
Balcony Falls
Potete's Gap
N

Alexander S. Paxton, a Washington College student, described the event as he witnessed it:

> A lot of the Unionists in the town decided to show their loyalty to the old flag by unfurling it from top of a lofty pole, to be raised in front of the court-house. Late one evening this pole was brought in, dressed and left lying in the street ready for raising on the morrow. That night some one by boring holes, made it useless for the jubilee. I never heard who did the mischief, but they blamed the cadets and indignation was at a boiling point. . . . a small party of cadets . . . were assaulted in a store, by a much larger number of town roughs. The cadets got the worst of the skirmish and beat a hasty retreat to the barracks, where they related their rough treatment. It was the spark, and the pent-up fires burst forth. Some of the professors, hearing the confusion, appeared on the scene, but they could not restrain the maddened crowd. Some, in their impetuosity, started at once for town with guns and ammunition, in squads, or straggling, as each one got ready. Captain [John] McCausland . . . told them if they were determined to go, to organize first and go with some system. This resulted in halting the advanced squads at the old tavern, at the foot of Main Street, about half a mile from the Institute. In the wagon yard they awaited the arrival of others who poured in as a turbid stream.
>
> About 3 o'clock in the evening, on looking out of my window in the College on the hill near by, I saw the gleam of bayonets. Rushing down to the tavern, I found the yard full of cadets in a state of wild excitement. Major Gilham, the commandant, told them to form companies, and it was but the work of a few minutes to organize. . . . Up town all was excitement, as the report spread that the cadets were coming on a mission of vegeance. The Union party were arming themselves to give the young soldiers a warm reception, knowing they had the advantage in firing on the unprotected lines from doors and windows . . . Down in the yard at the foot of the hill all things were about ready for the forward movement. The companies had their officers, their cartridge-boxes were full, and all were waiting for the order to march. Just then a tall military-looking man came up at a rapid pace, and, ascending the stile, stood with arms folded and for a moment surveyed the scene before him. It was Colonel Francis Smith, the superintendent, who it seemed had just heard of the trouble and came in the crisis of the affair. The moral grandeur of that scene surpassed anything of the kind I have ever witnessed. There he stood, calmly looking down into

the flashing eyes and glittering bayonets ready to be bathed in blood of Lexington's citizens. . . . Stretching out his hand over the battalion the buzzing gradually subsided and at length there was silence. Then he said, "Young gentlemen, I know you have received a great wrong, and you have my sympathy. I am your friend. This is not the way to right the matter. I appeal to your reason and better natures. A moral victory is better than a bloody one. Follow me and I will see that you get redress." Then stepping down from the stile he started for the Institute. There was a dead silence for a few moments, but moral influences triumphed and the crisis was over. One and another began to say, "Let's go back," and all of a sudden, as if by magic, broke ranks and all followed their esteemed instructor . . . . On arriving at the Institute, Colonel Smith took them into one of the society halls, gave them good advice, showed them their rashness and how near it led to a tragedy written in their own blood. . . . Some of the professors made speeches and the prospect of war was touched upon. It was then that Jackson, afterward "Stonewall," made the famous remark, "If we have to fight, let us draw the sword and throw away the scabbard." An order was issued that for the present no cadet should go up town.[14]

The townfolks' newly developed propensity for raising flags "was bound to infect the student body" of Washington College, wrote John Newton Lyle of that institution. Lyle added:

> More to tease Doctor [George] Junkin than to express political sentiment, some of the boys would climb the dizzy heights of the cupola of the main building, where stood the statue of Washington, to which they would fasten a staff bearing "The Bonnie Blue Flag with a single star," the emblem of South Carolina's sovereignty.[15]

Elizabeth Randolph Preston Allan was a 13 year-old witness to the disturbing events of the day. The daughter of Colonel J. T. L. Preston, a founder of V.M.I., recalled:

> Dr. Junkin was at breakfast when the janitor reported to him what had happened. He ordered the man to remove the flag, but the Negro knew better than to face that cheering mob.
>
> Finding his order not obeyed, the irate old man rushed over to the campus in dressing gown and slippers and called for silence. The habit of obedience held the boys quiet for a moment, while in a few stormy words, he reminded them that Virginia was still in the Union and that he was still President of the College.

"I order you to take that flag down," he cried.

But his brief opportunity was over; the boys yelled for the Southern Confederacy, sang "Dixie" and went mad with patriotic fervor.

Then to their surprise, the figure of my brother Willie, a seventeen-year-old freshman, appeared on the roof of the Colonade.

"Boys," he said, "Dr. Junkin is right; Virginia is still in the Union; and he is still our President. We must wait a few days, and I am going to take the flag down." Young Preston continued to climb toward the flag.

Pandemonium broke loose: the crowd stormed, cursed, threatened, and began to throw such missiles as could be found. Fortunately there were not many available on that smooth sward. The gallant climber went on, unharmed; reached the statue; took down the flag, waved it aloft and called for three cheers for the Southern Confederacy. They were given with a will; then he rolled the flag round its staff, thrust it into the bosom of his coat, and descended to an enthusiastic crowd.[16]

The next morning another flag was found flying from the roof of the building. Doctor Junkin had it removed. Each day a new flag appeared. "These performances took place at night," recalled Lyle, "and in the morning the Doctor, on discovering the banner, would order it taken down and burned by the college janitors. His flag burnings were conducted with all the solemnity of a state function."[17]

On April 17, the students petitioned the faculty for permission to fly the flag of the Confederacy atop the college. The request was signed by 61 students. Doctor Junkin regarded the petition as a personal insult and asserted ". . . that he would never give another Lecture or hear another recitation at the Institution until the flag was taken down; and, if the Faculty did not have it removed at once, he would call a meeting of the Board of Trustees, and hand in his resignation — that this was his *Ultimatum*. . ." The faculty allowed the flag to remain.[18]

President Lincoln's call for 75,000 troops to crush the rebellion in the South was the final blow to Unionist hopes in Virginia. Governor Letcher denied Secretary of War Simon Cameron's requisition for three regiments from the Old Dominion with the retort:

The militia of Virginia will not be furnished to the powers at Washington for any such use or purpose as they have in view. Your object is to subjugate the Southern States, and the requisition made upon me for such an object — an object in my judgment not within the purview of the Constitution or the Act of 1795 — will not be complied with. You have chosen to inaugurate civil war, and having done so we will meet you, in a spirit as determined, as the Administration has exhibited toward the South.

Despite Letcher's strongly worded refusal to supply troops, the governor staunchly rejected all suggestions that he seize federal military installations in the state until the convention voted for secession.[19]

CHAPTER II

"Old Rockbridge Is Thoroughly Aroused"

Doctor William S. White, pastor of the Presbyterian Church, reported the situation in Lexington at the time:

> . . . the 15th of April, 1861, I made a speech to a crowd of Union men, advocating their views. On the following day Lincoln's proclamation came, which I then regarded. . . as a declaration of war. Thus forced to fight, I claimed the poor right of choosing whom to fight. Necessity was laid upon me to *rebel* against *him* [Lincoln], or my native state. I chose the former, and became a rebel, but never a secessionist.[1]

James D. Davidson, a prominent Lexington lawyer and strong supporter of Letcher, wrote to Robert M. T. Hunter, a United States senator from Virginia:

> Allow me to say that I have been a Union man of the strictest faith until the Proclamation. On the 16th . . . I gathered around me in my office, some thirty of our strongest Union men, and prepared a dispatch to our delegate Dorman, in which they all concurred: "Vote an ordinance of Revolution at once." Rockbridge was revolutionised at once; and now our secessionist friends say they are the true conservatives and that we are the fire eaters. . . . Propositions, I understand, are now being made, by men of high position, looking to peace. Could they be sent by a Commission to the Southern Congress, looking for a peaceful separation? We can never be reunited again. The line is already marked between the cold north and the warm South, which cannot be wiped out, *even by blood.*[2]

Henry C. Davidson had been attending a fraternal meeting in Wheeling while the divisive fires were smoldering in Lexington. En route home, he had stopped in Washington, where he bought a copy of a daily newspaper and read Lincoln's proclamation. On April 18, he wrote:

> I left Virginia a Union man, but returned to it in favor of separation. I found the people of Virginia on the whole route from Alexandria to Lexington in a state of intense excitement & ready for revolution. Rockbridge, which was so intensely union, is now a unit for [illegible]. A special newspaper arrived here this morning for all the Volunteers of the County to march immediately — & orders the Col [colonel] of the Regt [militia] to

prepare his men for marching at an hours notice. Two Companies, the Rockbridge Rifles (Capt. [Samuel] Letcher) [and the] Rockbridge 1st Dragoons (Capt. White) from Lexington are now on their way to Staunton. The Rockbridge 2nd Dragoons (Capt. McNutt) from Brownsburg are also on the way & another company [from] the Forks region will march tomorrow . . . . Three fourths if not four fifths of the Volunteers gone are union men . . . . This is a terrible day for our country. It is hard to explain the policy of the administration. They could not have devised a surer plan of uniting the south than by the coercive policy they have adopted. God knows how it will end. Ruin — Ruin staring us in the face.[3]

Cadet Gatewood, ignoring the niceties of punctuation, reported on activities as viewed from his V.M.I. vantage point:

We received a *Dispatch from* Gov Letcher last night at two o'clock, wanting Col Smith to come to Richmond immediately, we started this morning at 8 o'clock in the Dispatch ordered the *Rockbridge Rifles* and a Cavalry Company to be in Staunton tomorrow morning by six o'clock the Rifle Company started today at one o'clock they passed the Institute and [we] saluted them, the Cavalry Company left this evening at 6 o'clock P.M. they are ordered to *Harpers' Ferry.* In the same Dispatch we received orders to hold ourselves in readiness and to drill 3 times a day. We have drilled 3 times today We have suspended all Academic duties except *Tactics.* All of us are studying *Tactics* We dont *do anything* at *all* but study *Tactics* and *drill*

Later Dispatch

We received a dispatch this evening at 3 o'clock saying, Harpers Ferry has been taken from the federal troops, Gosport Navy Yard had been taken, two million had been taken from a recked vessel, Virginia has seceded by a unanimous vote (Three cheers for Virginia) We fired a salute of 15 guns (cannon) this evening in honor of the glorious Old Dominion. We will be sent all over the state in a few days to drill recruits for the Southern Army . . . . There is a Cavalry Company out in front of barracks now after sabres nearly nine o'clock at night. What do you think of that. That looks like War. It is bound to come. . .[4]

A day later Cadet Thomas A. Stevenson penned this update:

Great excitement prevails here since the reception of the secession ordinance. . . . We drilled from five in the morning until dusk with slight intermissions. . . . The place about presents a most busy aspect — Arms are being carried from the arsenal armory the entire day to supply the Militia and Volunteers of the

County. Five companies equipped for service passed here en route for Washington. It was quite touching to see their wives and children following after. Or [our] entire second class are employed in making cartridges and tents are being repaired and every warlike preparation is being made.[5]

The *Valley Star* published the following news article on the stirring events in Lexington:

On Thursday last [April 8] at one o'clock, the Rockbridge Rifles (Sam'l H. Letcher, Captain,) took up their line of march, and in the evening the Rockbridge troops, (M. X. White, Captain,) took up their line of march and we understand were joined by Captain McNutt's Troop, of the Brownsburg District. Three other companies are being formed. The Rockbridge Grays, (Captain Updike,) numbering about 100 men, are quartered at the Va. Mil. Institute, where they are being drilled by the remaining corp[s] of cadets. An Artillery Company, now numbering about 75 men, is being formed by our citizens — and Profs. White & Nelson, of Washington College, are forming a volunteer company, and are daily drilling.

The above companies will hold themselves in readiness to march at a moments warning. On Sunday last 180 cadets left the Va. Mil. Inst. for Richmond. The school had been suspended at the Va. Mil. Inst. Many of the students of Washington College have gone into service. Dr. Junkin, the President, resigned on Thursday last, and his resignation was accepted.

The departure of the Rockbridge Rifles was described by the *Gazette* in this manner:

When all was ready, after bidding adieu to friends and relations, the Rifles were formed in front of the Court House, where Rev. Mr. Tebbs and the venerable Dr. McFarland stood prepared to call upon the Lords of Hosts for his protecting care under the trying circumstances in which they were soon to be placed. All heads were uncovered while the men of God appealed to the God of battle, to bless those who had left their wives and children, fathers and mothers, brothers and sisters, to go forth in defense of the rights bestowed upon us by our fathers of the revolution, and to protect and defend them from those who instigated by folly and wickedness, would deprive us of our dearly prized institutions. All hearts were softened, all eyes were moistened by the tear of sorrow for the necessitation of the case. Every soldier was determined, if need be, to stand to the last, and die in a cause so just and a service so honorable.

A similar service was performed for the Rockbridge 1st

Dragoons, by the venerable Dr. White.[6]

The young Elizabeth Preston remembered the departure of the V.M.I. Cadets for Richmond to become drillmasters for the rapidly gathering volunteers:

> Major T. J. Jackson was in command. They were to set off on their Sunday morning march to Staunton at twelve sharp, after a prayer from Major Jackson's pastor, Dr. White, for a blessing upon the young warriors. Fifteen minutes had been allowed for this prayer; the minister took only ten, and for five minutes of tense silence on the part of the ranks, their Commander sat like a statue on his horse until the Institute clock boomed the midday hour, when, at the last stroke, the sharp command rang out "Forward, march."[7]

At Washington College, meanwhile "the crisis came speedily and recitations ceased. A military company was formed, and the campus was converted into a parade ground. Squad drill was the order of the day. Professors and students kept step, marked time and double-quicked under the direction of drillmasters furnished by our sister institution, the V.M.I."[8]

John Newton Lyle reported that the college students hastened to join the company, which adopted the name of "Liberty Hall Volunteers," after a similar organization that had opposed British General Banastre Tarleton during the American Revolution.

Initially, the Liberty Hall Volunteers encountered parental objections to their martial plans. Sires protested against their sons marching off to war with fellow students serving as officers. This objection was removed quickly when Professor Alexander Nelson was elected captain of the company. Ill health forced Nelson to resign shortly thereafter. He was succeeded by Professor James J. White, son of the local Presbyterian divine. William Nelson Pendleton, West Point-trained and rector of the Lexington Episcopal Church in 1861, drilled the college company until he was elected captain of the Rockbridge Artillery.[9]

As the passions of war permeated the campuses of the V.M.I. and Washington College, the county fathers took measures in support of their youthful patriots. A meeting, chaired by Colonel James H. Paxton, was held at the courthouse, at which the following resolutions were adopted:

> 1st. Resolved. That a committee of seven men, one from each magisterial district, be appointed, who shall receive subscriptions of money and materials for clothing, consisting of Flannels, Jeans, Blankets &c.
>
> 2nd. Resolved. That J.[ohn] H. Myers be appointed Treasurer, to whom all monies shall be paid, which will be disbursed upon order of an Executive Committee, consisting of Hugh Barclay, Capt. George A. Baker, and William White.

3rd. Resolved. That R.[ichard] H. Catlett be appointed Quarter Master, to whom all materials for clothing shall be delivered to be manufactured into clothing suitable for soldiers.

4th. Resolved. That able bodied young men not enlisting . . . who desire to serve their country, be required to report to Colonel Davidson, at Lexington, immediately.

5th. Resolved. That Wm. Dold be appointed Commissary, who shall secure supplies of Flour, Meat and Vegetables, for the support of the soldiers awaiting orders in Lexington.

Those selected to solicit subscriptions in the various magisterial sectors included Eli S. Tutwiler, Lexington District; Major J. C. Davis, Natural Bridge District; Alexander M. Glasgow, Fairfield District; Colonel Samuel A. East, Brownsburg District; Colonel James H. Paxton, Paxton's School House District; Bolivar Leech, Collierstown District, and Major Samuel D. Gilmore, Kerr's Creek District.[10]

In other actions, the Lexington Town council asked the V.M.I. authorities to furnish 75-100 muskets to a Home Guard company being organized in the village. At the same time the faculty of Washington College turned over a quantity of old muskets to the mayor of Lexington.[11]

On April 27, Henry B. Jones of Brownsburg noted in his diary that a Home Guard company was formed in the village, with himself as captain, and four neighbors named Brown, Gilkeson, Walker and Hutcheson as lieutenants.

What was true of Brownsburg was also true in other Rockbridge County communities. "The whole country is a camp," wrote James D. Davidson from Lexington.[12]

Few, if any, in Rockbridge County expected the clash of arms between the North and South to be of long duration. Thomas McGuffin was one of them. Writing to his son in the Fifth Virginia Infantry, McGuffin commented:

> Putting all jest aside, John, I fear you will have to go [to Harpers Ferry]. How much I dread it. I entertain the hope that you will only have a brush & all will be restored to peace and harmony again. What a state of affairs. Our quiet Union, the envy of the old Countries, broken in pieces — First one corner & then another, scale by scale. Not by one unanimous & universal crash as if done by man blindfolded, but inch by inch here a little & there a little with their eyes open to its awful effect. I am not a great hand on "politics" but oh my, "war" calls forth my every sentiment & exerts my every nerve."[13]

As more military companies burgeoned throughout the county and war preparations escalated, the Rockbridge court authorized $25,000 in bonds to support them. The court also allotted $5 to clothe and equip each volunteer. Expenses were not to exceed $20 per man for quarters and subsistence, and transportation was limited to $5 per man. The court also was mindful of the folks who remained behind, appropriating $5,000 for the support of the families of soldiers.[14]

When Captain James White offered the services of the Liberty Hall Volunteers to the service of the state, Governor Letcher informed the former Latin professor that he could not accept the company because of the youthfulness of its members. The governor's posture imposed only a temporary setback on the hopes of the company, however. John Newton Lyle, the 22-year-old valedictorian of the Class of 1861 at Washington College — he never had an opportunity to deliver his valedictory address because of the intervention of hostilities — remembered:

> The attitude of the governor defeated our plans for equipment. All the other troops that had gone to the field were fitted out by the county. When we applied to the county commissioners, the treasury was closed to us. The kind old officials, however, almost shed tears when they explained that it would be highly improper to expend public money on a company that the governor had refused to accept.
>
> But fortune favored the brave and all things come to him who waits. The ladies of Rockbridge County, hearing of this action of the court, sent us word that they would equip and send us forth as their special knights to do battle for their dear old mother, Virginia. Here was glory without the smell of gunpowder. The fair sex had chosen us as their own champions, and our breasts puffed out with pride. And right royally did our ladies fair prepare us for the camp and march. Each lad was provided with everything a fond mother might dream her son might need, even to a needlebook, buttons and thread, linen gaiters, a havelock to screen his neck from the rays of the sun, and a red flannel waistband to wear next to the skin to keep off the diarrhea.
>
> They made our uniforms with their own hands. To do this work they employed a tailor to cut out, whilst they formed a circle and did the sewing in one of the public halls of the town. The Liberty Hall Volunteers were diligent in visiting this hall when not at drill on the campus. It was not the interest in the work so much as the presence of fair maidens there plying their needles that attracted them thither. When they tried on their uniforms, they found out that their visits were remembered. As a L.H.V. would haul on a pair of pants, he slapped himself, gave

a warwhoop and jumped to the ceiling of his room. Those mischievous girls had stuck pins in the garments. If they could have witnessed the performance, how they would have admired the agility of the boys.

Ah, those lassies of 1861, shall we again see their like. Even the thought of them makes the blood of old age gallop. They loved their native State with all their hearts, and were the best recruiting agents she had. If their reproving glances failed to make a swain enlist in her defense, they made life miserable for him. They would send him dolls and dollgags, refuse his escort, chide and rebuke him until, in sheer desperation over their cruel treatment, he joined the army. Those were gay times the L.H.V.'s were having playing soldier in Lexington, having no studies, no duties, and nought to do but drill an hour mornings and afternoons, and court the girls in the evenings. We marched to the war inspired by the tune of "The Girl I Left Behind Me."

The influence of the college and the Virginia Military Institute upon the society of Lexington had been the greatest benefit in the way of education and refinement. There was in its midst the extremes of neither poverty nor riches. The intense chase after the almighty dollar had not seized upon it. To integrity, good manners and virtues its doors stood open in summer and the latch string hung out in winter. What a contrast to the ease and elegance of such a community where our experiences of the four years of war upon which we were entering.[15]

In early May, Thomas McGuffin of Goshen Bridge addressed a letter to his son, serving with Company L of the Fifth Virginia Infantry at Harpers Ferry, in which he commented:

There is no news of interest about Goshen. The war and incidents connected thereto is the all absorbing Topic. There are several Volunteer Companies now drilling in the County, awaiting orders to March & making an aggregate of several hundred beside what has gone. Old Rockbridge is Thoroughly Aroused and prepared to Contribute a full share in resisting the Military Despotism now operating in Washington.[16]

At about the same time the *Valley Star* reported the formation of a Committee of Safety and a Home Guard company of 106 men at Goshen Depot. The "Rockbridge Grays," raised in the Buffalo Forge area, departed Lexington on May 3. En route to Harpers Ferry, the Grays were presented a handsome flag by the ladies of Harrisonburg. According to the *Star*, "This mark of respect, coming as it does from the fair daughters of a distant sister county should be highly prized by our gallant young men. All praise to the daughters of Harrisonburg."

In the same issue, the editors of the *Star* heaped laurels on the ladies of Lexington. For two weeks, reported the newspaper, the ladies "have been showing their devotion to the cause of the South. . . in making uniforms for the volunteers of Rockbridge. The men of the Artillery Company, under the command of the Rev. Mr. Pendleton. . . are nearly all uniformed, and they look well, and we think will do good service, when the time comes for action."[17]

Sectional acerbity took on a fraternal tone when the Welsh brothers exchanged sharp letters at a time when the chasm between North and South was growing ever wider.

John P. Welsh, a farmer in rural Rockbridge County, replied to a hostile letter from his brother, James L. Welsh, who had migrated to Illinois in 1853, in this manner:

> We have quite exciting times now here as I perceive you have there. You hear nothing but war, war, and to give you some conception of it, old Rockbridge has now five companies at headquarters and two more ready to go at a moment's warning. About 700 in all have gone and are ready to go and they are drilling men all over the county.
>
> So far as I can learn, the North expects to crush the South at once, but they are mistaken. We are prepared for a protracted war; we look upon it as a just cause and we are united men and women; all contribute without stint. The companies going from here and all parts of Virginia so far as I can learn, are fitted out and uniformed without costing the state anything; a people united and determined as the South is, won't be conquered in a day.
>
> I was very much pained to find that you had turned against your mother state; to think that I have a brother who would advocate sending men here to butcher his own friends and relatives and burn the house over his mother's head that reared him; but I can see that you have been imposed upon by false statements, else you would never have called your brother a "rebel." Old Lincoln will have to hang a goodly number of his colored breathren as they are almost to a man for resistance to him. Even the free niggers are volunteering to work on the fortifications at Harper's Ferry. . . . This is election day to say whether we will secede or not. I have always opposed secession but I shall vote for it today because I don't intend to submit to black Republican rule any longer. We proposed every term of adjustment that an honorable people could propose and were

willing to [do] anything to save blood, but no. The Republicans say, "You must submit, we have the lead now, we have the government in our own hands, we have the army and the navy, we will have no compromise, we will make you submit." Well, maybe they can since a man when he becomes a Republican forswears home, mother, father, and brothers and [is] willing to sacrifice all for the dear nigger. You talk of hanging Jeff Davis and others; just let me inform you that Virginia has 100,000 men now on duty and as many. . . ready to go, no hirelings are they either; men worth $40,000 go into the ranks with the common soldier. We don't intend to trouble the North but they must let us alone. Write and say what you please, it will make no difference to me. The same blood flows in our veins. You are as dear to me as ever, farewell. God be with the right and forgive the errings.[18]

Shortly after the letter was written the Welsh brothers entered the service of their respective armies.

On May 11 the Rockbridge Artillery marched out of Lexington. Before leaving, the 76-man company sent a letter to the *Gazette* in which gratitude was expressed to the ladies of the community for making its uniforms and to the townspeople for providing its equipment. James D. Davidson presented the company flag to Captain Pendleton "by command of the Ladies of Lexington — a command I must obey — I present in their name, to the Rockbridge Artillery, this flag — the work of their own hands."

Pendleton received the banner on behalf of the company and, after a few remarks, a formal acknowledgement was given by Sergeant James C. Davis, who appealed to the men to "never suffer this banner to be lowered in disgrace, or trail in dishonor before the minions of a tyrant."[19]

Equipment of the Home Guard companies suffered greatly when compared to that of the regularly enlisted units. A letter by Captain William W. Shields to the editor of the *Star* reflects that disparity:

> To give you some idea of how we are armed in this section of Virginia [in the Longwood area of Rockbridge]. . . I here give you a statement of the arms belonging to the Home Guard.
>
> Number of persons forming the Guard 98. Rifles 44, shotguns 38, a portion of them double barrel, 11 pistols, 2 muskets, 3 Bowie knives. PS: If you think proper you can publish the above. It may be consoling to our Northern friends to see that we are somewhat prepared to escort them through this region when that Big Army comes.[20]

For more than a month after the convention in Richmond voted to take Virginia out of the Union, the Old Dominion went its independent way. That situation changed on May 23. A statewide election — an election in which the outcome was a foregone conclusion — propelled Virginia into the Confederate States of America by an overwhelming majority.[21]

M. S. Roadcap reported the results of the plebiscite to his son. Zacheriah White of Lexington, he wrote, was the only man in Rockbridge County to vote for Virginia remaining in the Union. In addition, according to Roadcap, a gentleman in Bath County, Struck by name and a native of Pennsylvania, also cast a Union vote, an act for which "he was seized and well ducked and ordered to leave within 48 hours."[22]

Following the formal vote for secession, Governor Letcher issued a proclamation to the people of Virginia on June 3. In it, he called for all volunteer companies to report to previously designated rendezvous, requested more companies be organized and report, and required the militia to arm and prepare themselves for mobilization and to hold weekly drills.

The editor of the *Star* reported that Letcher's proclamation precluded men serving in the militia from volunteering in a company of their choice if called to active duty. He encouraged those subject to call to enlist in the "Rockbridge Rangers," a mounted unit then being organized in Lexington. The *Star* reported further that citizens of the county had heard the boom of cannon during an engagement at Aquia Creek, more than one hundred miles away.[23]

The orders that the Liberty Hall Volunteers had awaited for a seemingly interminable period finally arrived on June 3. The command was ordered to Harpers Ferry, where it was to serve under Colonel Thomas J. Jackson, its fellow townsman from the V.M.I.

June 8, the date for departure, found 74 officers and men eagerly awaiting the civic ceremonies that were necessary before starting their northward movement. As recounted by John Newton Lyle:

> The company formed on the college campus, immediately in front of the main building, marched into town and into line on Main Street opposite the county courthouse where a vast concourse of people were congregated. They had come from all parts of Rockbridge County as every neighborhood within its limits had a representative in our ranks. The whole population of Lexington, both white and black, were on the streets. It did us proud to see such a manifestation of our popularity. The company stood at parade-rest while a magnificent Confederate flag, the gift of the ladies of the Falling Springs congregation, was being presented. The Reverend John Miller, a native of New Jersey, made the presentation speech, and our Captain

White responded for the Liberty Hall Volunteers.... The motto it bore was "Pro aris et focis," which translated is "For your altars and friends."

Alexander S. Paxton, a 22-year-old member of the company, recalled that: "Captain White, in quivering accents, accepted it [the flag]. There were few dry eyes in that great crowd who came to see us off."

Lieutenant Lyle picked up the narrative:

> Captain White's father, the Reverend William S. White, D.D., pastor of the Presbyterian church, invoked upon us the blessings of the God of our fathers. Women and children were crying and tears were coursing down the cheeks of bearded men.... With heartstrings ready to snap mothers and sisters pressed sons and brothers in farewell embraces as they bid them go and do battle for Old Virginia. Those women were descended from that heroic stock that, amid the moors and mossledge of Scotland, had sacrificed their all for God and Presbytery. And no less, but louder than theirs, were the lamentations of the black mammies who had come to say goodbye to their "chil'n" now grown to be their young masters, and press them to their warm hearts.... The sweethearts of some of the young soldiers were present and wept silently behind veils.

Alexander S. Paxton concluded the account of that historic exodus;

> The drum beat and at the command, "Forward, march," we went to the foot of the hill, where, getting into a long line of stage coaches and hacks, we tearfully took leave of dear old Lexington and our alma mater.[24]

As the Liberty Hall Volunteers headed for its date with battlefield destiny, another Rockbridge County company, the "Fairfield Rifles," was taking form. According to an announcement in the *Valley Star*:

> This Company, already numbering over 50 members, is rapidly filling up, and will be organ. in a few days. Young men desiring to join this company should report themselves immediately, otherwise they may be drafted in the militia. We hope this company may, by the end of the week, number 75 or 80 men, and organize immediately and proceed to Staunton. We want to see that section well represented in defending our country.

The same issue of the *Star* contained an account of the Battle of Phillippi (June 3), in which the Second Rockbridge Dragoons played a part.[25]

Elizabeth Randolph Preston Allan, a youthful witness to all the excitement and tearful farewells, remembered the dramatic impact of war on the populace of Lexington:

> How speedy was the evacuation of Lexington on the part of all men available for military service. . . . the little town we knew, once overflowing with young men in peaceful pursuit of learning, now beholding its male population reduced to the aged and infirm.
>
> The women of Lexington promptly resolved themselves into sewing bands, to prepare hospital supplies as well as additional shirts, etc. for our hastily equipped soldiers. Old linen was brought out to be scraped in piles of soft, snowly lint for wounds; bolts of cotton were quilted; pillowcases made, and sheets, as well as blankets were contributed from every house keeper's store. Before the war was over our carpets were taken up, cleaned and used for hospital coverings, or to keep our men from freezing in their camps, tents and block houses.
>
> Much time was wasted in making what was called "Havelocks" — a brown linen covering for the cap. . . . our men found them cumbersome and discarded them. I think, however, that the haversacks we made were used to carry food on forced marches, and perhaps the covering we made for the tin water canteens which the soldiers slung over their shoulders. At first these activities took place in the churches, Sunday Schools and vestry rooms, but as these places must continually be put back into shape for other uses, eventually the women took possession of a large hall belonging to the Masons.[26]

Meanwhile, the ladies of Brownsburg were equally as busy in preparing a local company for the vicissitudes of war. A letter to the editor of the *Gazette* contained the following:

> The Rockbridge Guards under the command of Captain David P. Curry, left this place for Staunton. . . the 5th . . . . The company numbered seventy five rank and file. . . . The company is made up of most excellent material, every man in it is young, hale and stout, and I venture to say it is one amongst the very best companies that have left our county for the seat of war and being under the command of an intelligent Captain and other officers, and having been well drilled for some time before leaving by Col. A.[lphonso] Smith of Lexington, I have every reason to believe they will render good service to their country. I suppose there has no company left the county better, or more thoroughly equipped, every man in complete uniform, with coats and pants of gray cloth, and most of them with two fatigue shirts, knapsacks, haversacks, canteens, bed sacks, and double tents — with new shoes and socks for all. In a little

over a week, the ladies of Brownsburg and vicinity and on North [now Maury] River, made 80 coats, 80 pants, about 140 fatigue shirts, 80 knapsacks, 80 haversacks, 80 cloth caps, and covered 80 canteens with cloth, and ten tents.

They deserve, and I have no doubt will receive the lasting honors, and praise of the county for their indefatigable industry in fitting out this company. On the even of their departure, a very fervent prayer was offered in their behalf by the Reverend E.[benezer] D. Junkin and an appropriate address by the Rev. W.[illiam] W. Trimble, which was responded to in a very feeling manner by the Captain.[27]

In early June the county was thrown into a state of near panic by word that a Federal force had advanced as far as Warm Springs in Bath County and was headed for Rockbridge. "Get up, get up, the Yankees are upon us," cried a courier as he raced through the streets of Lexington, telling the men to rally on the courthouse green.

A Brownsburg correspondent of a local paper reported:

From 10 o'clock to 4, there was collected at this place 75 or 80 men, at J. W. Youell's, on Walker's Creek, 55, with gun in hand, and about 10 o'clock the same night 42 men from New Port and vicinity made their appearance here, all ready to march to Lexington at a moments notice.

I have understood that about 100 had met at Fairfield, and 50 at Timber Ridge for the same purpose.

Francis T. Anderson reported the situation in the lower end of the county:

We had quite an excitement yesterday in the Valley. Early after breakfast I received a letter informing me that a dispatch had been received from Frank Shields that the enemy in force, 11,000 strong, had taken possession of Lewisburg. I immediately took measures to call the militia together. A few hours afterwards I received a message. . .that there were 1,300 Northern troopers within a short distance of Lexington and were expected to reach there by twelve or one o'clock. I immediately sent out runners and by one o'clock had a force of sixty able bodied at the Ferry [Greenlee's] and about 40 negro men ready to march to the relief of Lexington. We crossed the river, and I rode on in advance and met Mr. Paxton who informed me that it was a false alarm.[28]

A reporter for the *Gazette* recounted the plight of the Second Rockbridge Dragoons:

> This company, since its connection with the Northwestern army, has, perhaps, rendered more perilous service for the State than any company in service — scouting night and day in the enemy's country — rescuing persons in jail for their adherence to the ordinance of secession — expeditions at night of 20 to 30 miles for powder and lead belonging to the enemy — an almost unceasing watch upon the enemy's scouts and spies, except the night before the attack at Phillippi, at which time they were worn out and not upon duty. Their encampment being always placed nearest the enemy, the attack . . . was first made upon them. They were the last to leave town.
>
> The Quartermaster of this company is now in this county for the purpose of getting Camp Equipage and clothing for the men, (most of whom lost all they had except what was on their persons.) Nearly all their arms were saved. The Quartermaster is authorized to receive as volunteers for said Company 15 or 20 brave men and fearless riders. They will be received without uniform. All that is necessary is to report to A. B. Mackey or C. W. Johnson, Lexington, before Saturday next, with a good horse and a double-barrel shot gun.[29]

Colonel Francis H. Smith of the Virginia Military Institute issued a call about this time for volunteer companies to join the "Military Institute Battalion," to be armed, equipped and trained at the Institute. The proposed Battalion, which was to include two artillery companies as well as infantry, may have been the "Rockbridge Battalion" that Captain William Nelson Pendleton started to recruit at this period. Neither plan reached fruition, although the volunteer companies continued to be armed and equipped from the arsenal on the V.M.I. campus. Students from the University of Virginia and other colleges also were sent to the Institute for officer training.[30]

At regular intervals, the *Gazette* kept its readers informed on the state of affairs with the "Rockbridge Mounted Rifles" being formed in Lexington. The uniforms, it said, were to be:

> Gray Sack Coat, Gray Shirt and Pants, and Drab M_____ Hat. . . . A lady in the county gives a Horse, Saddle, and Bridle, and a Recruit. We trust the men will not be excelled by the women in their liberality. Persons desirous of aiding the Company in the purchases of horses, etc., can do so upon application in Lexington. The great want of the service is unafraid men, let Rockbridge send 100 men well equipped. The company intends to join the Brigade of [ex] Gov. [Henry] Wise, now forming, and the enlistments will be for 3, 6, 9 or 12 months. Persons desirous of uniting with this company should report at once, as

they commence drilling in a few days.

Many of the men enlisting in this company were past military age. This troop, 80 strong, joined General Wise on July 4.[31]

By common consent the efforts by the ladies of Rockbridge County in equipping troops in the field were beyond compare. They were to be commended highly for their patriotic endeavors. However, an urgent need still existed. This pertained to hospital supplies. As explained by an unnamed female letter writer from Augusta County in the *Gazette:*

> They are sadly needed in the Northwest and at Manassas Junction. . . . I think we ought to turn our attention now to our army at Laurel Hill (12 miles from Phillippi.) But get everybody to work. . . . Sheets, pillows, ticks, (to be filled with straw,) pillow cases, towels, lint, bandages, rags, anything and everything — money for arrow root, medicines, &c., &c.

The editor added: "Ladies, you have done much for the well soldier — forget him not when he is sick."

When the supplies were accumulated they were to be forwarded to Staunton. In the next edition of the paper, the editor advised that the ladies in the Forks were preparing a box of supplies for the hospitals. He added:

> A few dollars from the gentlemen to buy cheap materials to make up would be very gratefully received. The patriotic ladies of that neighborhood complain that they have not been called upon to do half enough for our gallant soldiers, only give them an opportunity by sending them work to do, and see how the needles will fly.[32]

The July 7 edition of the *Gazette* published a letter from the members of the Second Rockbridge Dragoons. The communique expressed sympathy and respect for Lieutenant Robert McChesney who fell in battle on June 29, thereby becoming the first resident of Rockbridge County to make the supreme sacrifice in the war. The letter said in part, "we . . . bear testimony to his gallant bearing as an officer, soldier and gentleman."[33]

The death of the officer was but the harbinger of more sad news as the weeks went by. The Confederate defeat at Phillippi was followed by the debacle at Rich Mountain in which three of the Rockbridge Guards were killed, four wounded, including Captain Curry, and several captured. The First Battle of Manassas on July 21 delivered a crushing blow to the Upper Valley. Relatives and friends mourned the loss of 12 from the county. Thirty-five suffered wounds.[34]

The glamor of war, scarcely three months old, was quickly giving way to harsh, uncompromising reality. More troops were needed. Acting at the request of President Jefferson Davis, Governor John Letcher sent out a call for 3,000 more soldiers from Virginia. Initially, Letcher mobilized the militia, but it was allowed to return home when the Fairfield McDowell Guards, the Kerr's Creek Confederates and Captain Thomas H. Watkins company from the Natural Bridge area were sworn into Confederate service at Staunton.[35]

Military personnel and civilians eagerly devoured an article published in the post-Manassas period by the *Gazette*, now the only newspaper in the county and reduced to a single page. The invention of a large seven-shot revolver by a Mr. Keesee of the Rockbridge foundry on Irish Creek,

> ... is made of the Colt pattern, but in material is far superior to Colt's Navy Pistol, having this advantage: The cylinder being made of steel and the portances, which in Colt are of Cast Iron, are of wrought iron in Keesee's, another advantage over Colt's is that it has the mould for the Minnie ball, which gives it a decided preference.

The journalist expressed the hope that the revolver would be patented by the Confederate government. Another Rockbridge inventor, Henry T. Hartman, received a patent for a portable breastwork.[36]

Of greater importance to the Confederacy than these inventions were the iron furnaces and foundries of Rockbridge County. Because of the war demands, production schedules were stepped up at Jordan's, or Buena Vista, furnace, Glenwood furnace in Arnold's Valley and California furnace near Goshen. So that the skills of colliers, furnace workers and and blacksmiths could be utilized fully, some men were deferred or detailed from military service. Iron foundries were located in Vesuvius, Rockbridge Baths and on Irish Creek, all of which manufactured horseshoes, cannon balls and other iron products so vital to the Confederate war effort.

Other exemptions were granted to employees of the Monmouth Cloth Factory as well as to those of Thomas H. Deaver, who expanded his shoe factory to the point where it overflowed into the Odd Fellows hall in Lexington. The means of transporting the finished goods to the Northwest was via teams furnished by local farmers.[37]

Colonel Francis H. Smith had other matters on his mind. The superintendent of the V.M.I. appealed to the president of the Board of Visitors for his support in reopening the school in the fall. "In the meantime," he said, "the employees... under the charge of two of its officers, still on duty, might continue the musket cartridge laboratory now in operation, and which turns out for the use of the Confederate service 10,000 cartridges a day."

By September 2, the arsenal had issued 61,594 sidearms and 186 guns. Later the arsenal became a Confederate Ordnance Depot. By mid-December, 40 operatives were engaged in making cartridges.[38]

All the while the editor of the *Gazette* redoubled his pleas for hospital supplies. He recommended the formation of a Soldiers Aid Society, similar to those organized in other communities. He suggested further that readers begin collecting winter clothing for men in the field.

Reaction to these suggestions was instantaneous. Citizens, especially the women, were quick to respond. Mrs. C. C. Baldwin of the Forks community reminded the newsman that the Soldiers Aid Society had been organized on July 9. Moreover, she continued, "The Society in. . . July sent a box of hospital supplies to Manassas and one to Charlottesville, worth at least $100, and also ten dollars in money to the latter place, for the relief of wounded Rockbridge soldiers. . . . The Society expect[s] to get a box of supplies every month. They solicit contributions of all kinds, and particularly of wool rolls of yarn to knit into socks for our soldiers." The Ladies' Aid Society further requested that invalid soldiers be sent to Rockbridge County to recuperate in private homes. The Lexington Aid Society asked that blankets be sent to them as soon as possible and left at the Ann Smith Academy for girls, where they were to be bailed and forwarded to camps and hospitals. Seventy women and children were united in this enterprise.

Robert S. Campbell, the Commissioner of Revenue for the southwest district of Rockbridge County, rated an editorial salute from the *Gazette* for his efforts in rounding up supplies for the hospitals. According to the weekly publication:

> a gentleman more than seventy years old is busily engaged in collecting articles for the comfort of our soldiers. . . .He has fitted up a one-horse wagon, with which he collected the past week supplies to the value of about seventy-five dollars in the form of Hams, Butter, Lard, Eggs, Crackers &c. Almost every stage that leaves the town carries with it some of his collections.
>
> He remarks that the people of the county are remarkable liberal — ready to share everything they have with our suffering soldiers.[39]

James L. Campbell of Washington College served as the local agent for the Soldiers Aid Society. He filed the following report on his activities:

> Fifteen boxes (some of them very large) containing bedding, clothing of every variety, teas, sugars, crackers, and other eatables, table ware of different sorts, &c. . . . also two boxes of bottles, containing wines and cordials, have been packed and forwarded to several of the Hospitals, from the "Soldiers Aid

Society" of Lexington.

Three large and very valuable boxes, filled with sundry articles similar to those above mentioned, were sent by me to the Charlottesville Hospitals, from the ladies in the neighborhood of Fancy Hill.

Several packages of the same character were collected in the section of the county near Buena Vista Furnace, and forwarded to the North West.

Collierstown and Buffalo each sent a large and valuable box to Charlottesville, filled with the same variety of important donations.

The kindness and care of Mr. Clowes, the stage-agent, of the stage-drivers, and R.R. agents are worthy of all praise. . . . Supplies for the University [hospital] may be delivered at any point on the canal or Rail-road. They might be sent by wagons going to Millboro, from every part of the county.[40]

Despite the rising war fever and the drain on the area's young manhood, Rockbridge County remained a crucible of education in the Fall of 1861.

Lexington High School opened on September 1 with F. M. Edwards as principal. A Classic and English School was conducted at Ben Salem Church under Reuben Lewis, starting September 3. Mrs. Louisa Bull's School started classes on September 10, followed by the 54-year-old Ann Smith Academy, under William N. Page, on September 16.

The Brownsburg Female Academy opened on September 30, with the Reverend W. W. Trimble as principal. Charles P. Estill launched his preparatory school in the old "President's House" on College Hill on October 1. Professor John Koerber, formerly of the Virginia Female Institute in Staunton, opened a music school in Lexington during October.

Washington College started classes on October 6. Reigning fatherlike over all these institutions was the Franklin Society. Members of the organization, whose origins were believed to go back to 1796, met regularly in their hall at the corner of Jefferson and Nelson Streets to debate the most popular issues of the day.[41]

With battlefield casualties in their midst, and some in newly dug graves, members of the Ladies Aid Society of the Natural Bridge District decided the time had come to memoralize the deeds of valor by Rockbridge soldiers. The women proposed the erection of a monument on the Washington College campus to the memory of those soldiers slain at First Manassas. The women hoped to raise $5,000 for the project, which got away to a fast start when a Society member, Mrs. Aurelia R. Salling, donated $100 toward the memoial. Subsequently, the Liberty Hall Volunteers collected $70 and the ladies in the Forks area raised about $200.[42]

Little military activity occurred in Virginia during the fall of 1861. There was, however, no shortage of war news in the newspapers of the Valley. The major portion of column space was devoted to casualty reports, soldiers' letters that, in many cases, paid tribute to their fallen comrades; politics, and the plight of the sick and wounded in hospitals. Advertising dropped to a scant half-page and much of it contained requests for supplies, horses and teamsters for the army.[43]

To accommodate the wounded and diseased soldiers, a hospital was established at the Rockbridge Alum Springs. Chaplain C. H. Ryland reported 600 patients were bedded there. When the hospital quartermaster advertised for stoves and firewood to heat the wards, which had been converted from the outbuildings of the spa, the Ladies Aid Societies came to his rescue.[44]

After several months of relative battlefield inactivity between Union and Confederate forces, Rockbridge County residents had a new topic for discussion when the battle of Alleghany Mountain was fought on December 13. The Second Rockbridge Artillery, formerly the Fairfield McDowell Guards, suffered its first casualties in this engagement. Captain Watkins' company, now a part of the 52nd Virginia Infantry, and the Rockbridge Guards, a company in the 25th Virginia Infantry, participated in this Confederate victory.

Andrew Patterson wrote to his brother from Brownsburg about the battle: "Two of the Companies in the fight were from this end of the County — Capt. Currys — commanded by Lt. Whitmore, and the McDowell Guards, commanded by the Rev.-Capt. John Miller. . . . both companies fought with great bravery and Lieut. Whitmore, who is not much larger than your fist, charged upon the enemy up a hill against *double* his men and completely routed them."[45]

As the first year of the war drew to a close, many letters of reflection were written to the folks at home. One such epistle was penned by Benjamin Franklin Templeton, a member of the Second Rockbridge Artillery, to his children as he left for active duty. Templeton was 28, a "handsome and promising young man," who had returned to his native Rockbridge County at the outbreak of hostilities. He wrote:

> Now as I am going away on a dangerous journey for your defense, I think it but proper to leave on this page some little advice to you, & first obey your Mother. don't distress her by not doing what she tells you, or by doing what she tells you not to do — learn to read your Book that you may read your Bible, so

that you can better obey its precepts — & while your Father is absent from your hearth, pray for him every night, when you say your prayers — try and not get sleepy before you have said your prayers — when you[r] Mother say[s] when you have finished your prayer — say & bless, oh Lord my papa wherever he is this night and keep him from harm — and you Maria, say, may Pa rest safe this night — & now my children if you mind to do this, you[r] dear Pa will come back safe to you again — love to pray for you[r] Pa and *God will hear your prayers* — your [God] tells you this & does he not always tell the truth. Good bye.

Within four months after penning this poignant letter, Templeton died in a Staunton hospital, a victim of typhoid fever.[46]

As the year drew to a close, there was little to remind Rockbridge County citizens of mirthful holiday seasons before Mars raised his forbidding features. Elizabeth Randolph Preston, just turned 13, remembered:

> Sewing and knitting for the soldiers, packing boxes to send to the army, a quiet Christmas with ineffectual attempts on the part of the elders to be gay, for the childrens sake.

"And so the year '61 passed, with its excitements, its hopes and fears," wrote James D. Davidson.[47]

CHAPTER III

"Costly Victories"

The new year dawned with great promise for the Confederacy. Rockbridge citizens continued to lend their wholehearted support to the Southern cause and numerous letters appeared in the *Gazette* expressing gratitude for the people's generous donations to servicemen in camps and hospitals.

The editor of the *Gazette* also reported that quite a number of men of military age seemed to be walking the streets of Lexington while their peers were in uniform. The paper began publishing a list of names of men absent from duty. Originally, the list was a mere trickle, but it grew ominously as the war progressed and desertions became a major factor in the ultimate Confederate defeat.[1]

William Nelson Pendleton, now holding the rank of colonel, was still trying to raise companies for his proposed Rockbridge Battalion of artillery in January 1862, and Colonel Alphonso Smith of the militia was bending every effort to raise a heavy artillery company for service at Craney Island near Norfolk. As an inducement for enlistment, Smith offered each recruit a $50 bounty.[2]

Confederate setbacks at Fishing Creek, Forts Henry and Donelson, and Roanoke Island prompted the government in Richmond to increase its demands for more troops. On February 5, Governor John Letcher made his presentation to the lawmakers, explaining that some of the new levies would replace those men who had enlisted for one year, while the others would be applied to Virginia's quota in the army.

Three days later, on February 8, the legislators enacted into law a requirement that all males aged 18 to 45 must enter the army or enlist in the militia. On February 10, Letcher was empowered to draft militiamen in sufficient numbers to meet local recruiting quotas.

In other legislation, the governor was enabled to require county exemption boards to examine all eligible men with the authority to grant exemptions when they were justified.

From all indications, the Rockbridge board acted justly. It met at the courthouse in Lexington, March 12-19 and examined hundreds of adult men. Reconvening on March 22 and April 14, the board granted military exemptions to employees of the Jordan [Buena Vista] and Echols [Glenwood] furnaces.[3]

The county militia, numbering in its ranks the editor of the *Gazette,* marched to Staunton on March 20 where they were organized into three companies. The new editor of the *Gazette,* Cyrus H. Burgess, learned that 150 men of the Eighth Regiment were in the column, and 207 others were left behind because they failed to receive ample notice of their departure time. The record of the 144th Regiment was considerably better. About 200 reported, leaving behind 100 tardily notified members. The reason for the short notice was a call from General Thomas J. (Stonewall) Jackson, who was contending with a superior Federal force in the lower Valley.[4]

Jackson suffered his only defeat as a commander at Kernstown on March 23 and took large numbers of casualties among Rockbridge troops, especially the First Rockbridge Artillery. In the wake of the battle, Jackson's small army retreated up the Valley Turnpike to Rude's Hill, where it was augmented by the militia troops. As a result the remainder of the Rockbridge militia was sent home. And when the War Department refused to acknowledge the Rockbridge Battalion, its 250 recruits were allowed to join units of their choice.[5]

On the homefront, meanwhile, folks were bombarded with requests for scrap iron, lead, copper and brass. Bells were needed for melting down into cannon. The editor of the *Gazette* exhorted his readers to plant corn and to stop distilling it for liquor. Rather, he urged, it should be used to feed more hogs. Farmers also were encouraged to grow flax and sorghum. In other action the county court appropriated $10,000 for the benefit of soldiers' families, the money to be distributed in amounts of $20 or less. Robert I. White was appointed county agent and directed to purchase 10,000 bushels of salt which were to be distributed at cost, plus a reasonable commission.[6]

The exigencies of war were starting to take their toll on county households. Margaret Junkin Preston, wife of Colonel J. T. L. Preston and stepmother of Elizabeth Randolph Preston, confided to her diary.

> I loathe the word — War. It is destroying and paralyzing all before it. . . . all the able-bodied men gone — stores shut up, or only here and there one open; goods not to be bought, or so exorbitant that we are obliged to do without. I actually dressed my baby all winter in calico dresses made out of the lining of an old dressing-gown; and G[eorge] in clothes concocted out of old castaways. As to myself, I rigidly abstained from getting a single article of dress in the entire year, except shoes and stockings. Calico is not to be had; a few pieces had been offered at 40 cents per yard. Course, unbleached cottons are very occasionally to be met with, and are caught up eagerly at 40 cents per yard. Such material as we used to give ninepence for (common blue twill) is a bargain now at 40 cents, and then of a

very inferior quality. Soda, if to be had at all is, 75 cents per lb. Coffee is not to be bought. We have some on hand, and for eight months have drunk a poor mixture, half wheat, half coffee. Many persons have nothing but wheat and rye.

Mrs. Preston also noted that Washington College had only five students, all too young for military service. Snow was on the ground as late as April 10, 1862. On that date she added:

> For months we have had no service at night in any church in town, owing to the scarcity of candles, or rather to *save* lights and fuel. Common brown sugar, too dark to use in coffee, sells here now for 25 cents per lb. Salt is 50 cents per quart in Richmond. . . .
>
> One thing surprises me very much in the progress of this war; and I think it is a matter of general surprise — *the entire quietness and subordination of the negroes.* We have slept all winter with the doors of our house, outside and inside, all unlocked. . . and some $600 worth of silver, most of it in an unlocked closet, is in the dining room. Would I get my Northern friends to believe that?[7]

The arrival of a 400-man detachment from Lynchburg to join Jackson's army created a stir in Lexington. The editor of the *Gazette* complained that the men had arrived by canal boat without prior notification to the inhabitants of the town. After negotiations, permission was obtained to house the contingent in the buildings of Washington College. Moreover, the men carried no provisions, cooking utensils, candles or firewood. Once more Lexingtonians opened their hearts, furnishing the items to the soldiers before departure next morning.[8]

A skirmish near Williamsville in Bath County on April 26 provoked an air of rejoicing in Lexington. In the engagement, the Bath Cavalry attacked a Union forage train, destroying the wagons and bringing off 95 horses and mules and eight prisoners. The captured Yankees were lodged in the Rockbridge County jail until arrangements were completed to send them elsewhere.[9]

On May 1 Cadets of the Virginia Military Institute marched to Staunton where they joined Jackson's Valley Army. The battalion was commanded by Major Scott Shipp. Colonels Francis H. Smith and J. T. L. Preston also accompanied the Cadets, who marched in support of Jackson's forces but saw no action in the Battle of McDowell on May 8. Three Rockbridge units in the 25th, 52nd and 58th Virginia Infantry regiments, saw action in this fight, the sounds of which were clearly audible to Margaret Preston back in Lexington.[10]

In the spring of 1862, because of the scarcity of money in low denominations, the Bank of Rockbridge started to issue notes in the amount of $1.00, $1.25 and $1.50. Later the bank warned its customers to beware of counterfeit $2 bills being circulated in the county.[11]

In addition to the shortage of commodities already noted, the dwindling supply of leather was becoming a major concern to residents of the Upper Valley. One citizen complained to the *Gazette*:

> . . . the price has been raised to *one dollar and twenty-five cents* per pound, at the principal Yard in Lexington — an increase of *twenty-five cents on the pound*! The Hides cost from four to six cents — Leather $1.25!! Our soldiers are greatly in want of Shoes and Boots. Some of them in one of our nearest Camps are in their "stocking feet." If they get Boots they have to pay $14 to $15 for them — Shoes in proportion. Their pay is $11 *per month* — Looking at the comparative prices of Hides and Leather, I respectfully ask, can this accord with a good conscience? Can it be properly looked upon as anything but extortion?

The editor asked the women of the Soldiers' Aid Society to visit the families of absent soldiers in a friendly, warm-hearted way, to find out the real needs, and supply their wants as nearly as possible. He cautioned, "Those tokens of regard and sympathy must come from the *heart*, are often far more acceptable than those that come from the *hands*."

William S. White, the local Presbyterian clergyman, complained that on the day of national prayer all stores and shops were closed except the government wagon shops. To him, "This seemed. . . a strange incongruity to have the clatter of hammers and axes kept up within hearing of the congregation, who were requested *by the Government* to suspend business, and repair to the sacred temple of worship, to ask the blessing of Heaven upon our cause."[12]

The battles of Williamsburg (May 5) and Winchester (May 25) produced more long casualty lists among the servicemen of Rockbridge County. The "Valley Regulators," raised along the border of Rockbridge and Botetourt counties, fought with the 11th Virginia Infantry in the engagement on the Peninsula, while the Rockbridge Artillery suffered heavy losses in Stonewall Jackson's rout of General N. P. Banks in Frederick County.

One of the Winchester casualties was Frank Preston, son of Colonel J. T. L. Preston, who received news of the wounding when he returned to Lexington with the V.M.I. cadets. According to Mrs. Preston: "In about two hours, the carriage was ready, and Mr. P. on his way to Staunton. Prof. Nelson went with him, as his brother-in-law is slightly wounded."

In Staunton, Preston procured an ambulance and continued to Winchester where he learned that Frank's arm had been amputated at the elbow by a Confederate surgeon. When Jackson retired up the Valley, Colonel Preston was forced to flee his son's bedside to escape capture. During the Federal occupation of Winchester, a Union surgeon found that Frank's first operation was unsuccessful. A second operation was performed, removing the remaining portion of the arm at the shoulder. Frank convalesced in the home of a comrade in Winchester. His parents received no word of him until he walked into their home on July 21.[13]

The battles of Cross Keys and Port Republic (June 8 and 9) sent another wave of sorrow through soldiers' homes in Rockbridge. Among those slain was Alphonso Smith, the one-time Lexington officer who was fighting as a private with the 27th Virginia Infantry. Many more lay dying in hospitals across the state. One family's grief plunged to inordinate depths when a mother and three soldier sons expired within the span of three weeks.[14]

As deaths mounted, the *Gazette* published more and more advertisements in the settlement of estates. Frequently, slaves were included in such auctions. The estate sale of Sally Fulton advertised "a negro WOMAN and three CHILDREN, one MAN and one GIRL. These negroes are all young and lively and I believe honest and trusty."

In another newspaper advertisement, any man 45 years of age or older who was willing to serve as a substitute for a man of military age was offered 50 acres of land and $500. Whether the blandishment by an obviously affluent individual was accepted is not recorded.

As the war shifted temporarily from the Shenandoah Valley, desertions gnawed at the muster rolls of Rockbridge units. When the captain of the "Rockbridge Greys" submitted a list of absentees to the *Gazette,* the editor exorted the 14 men to report to duty instantly.[15]

A hospital established in the buildings of Washington College under the direction of Dr. Robert L. Madison of the V.M.I., evoked from the editor of the town weekly the wish that "our Ladies will go to work at once, in preparing changes of underclothing and bed-linen, to be on hand when the patients arrive, also that they will hold themselves in readiness to furnish good bread (the most important article of diet,) and other things suitable for the sick."

The faculty of the college took a dim view of the hospital in its midst and asked the Surgeon General to have the facility moved elsewhere. The request was granted. The Ladies Aid Society of Natural Bridge repeated an earlier offer to convalesce 100 wounded in the homes of its members.[16]

For several weeks in the summer of 1862 Rockbridge residents were forced to obtain war news through word of mouth or letters from servicemen. The *Gazette* ceased publication due to a shortage of paper, rags with which to make paper, or money to purchase it. It resumed publication on August 7.[17]

The early editions following the interruption contained news of the Seven Days battles around Richmond and reported the valiant conduct of the "Letcher Artillery," commanded by Captain Greenlee Davidson of Rockbridge County. Another news article reported the capture of the Union garrison at Summerville in Nicholas County by a Confederate cavalry force that included Captain John A. Gibson's Second Rockbridge Dragoons and Captain William A. Lackey's Valley Rangers.[18]

On July 7 the County Court appointed a committee, consisting of Jacob M. Ruff, William R. Moore, John M. D. Alexander and William W. Shields "to proceed to the battlefields near Richmond to inquire after the soldiers from this county, and if necessary, to bring home the wounded and sick; and it is ordered that their expenses be paid out of the money appropriated for the soldiers of the county."[19]

In addition to stories on war developments, the *Gazette* printed a plea by Captain Thomas H. Tutwiler, the quartermaster of Lexington, for farmers to sell him their hay and grain. At the same time, Lieutenant Colonel Samuel D. Baker of the Eighth Regiment of militia ordered all men between the ages of 18 and 35 to report to the Provost Marshal at Lexington for reexamination, regardless of prior exemptions. He also directed that all militiamen between 35 and 45 to report on August 16 so they could be reorganized.[20]

The battle of Cedar Mountain (August 9) produced more heartbreak for Rockbridge families, but the number of casualties among county men was relatively small when compared to those that followed the battles of Second Manassas and Sharpsburg.

For the second time Colonel J. T. L. Preston was called to the site of conflict. He was summoned to Northern Virginia following the battle of Second Manassas by word that his young son, Willie, had been mortally wounded while fighting with the Liberty Hall Volunteers in the Fourth Virginia Infantry Regiment. The colonel found the remains wrapped in a blanket in an unmarked grave. Identification was possible only by Willie's name on his shirt.

Tragedy for the Prestons did not end there. In December, 17-year-old Randolph, a Cadet at the V.M.I., died of typhoid fever.[21]

Caring for hospitalized soldiers was an ongoing project for the ladies of Lexington. Members of the Society of the Southern Cross sponsored concerts at the Methodist Church to raise money for food and clothing. The congregation of the New Providence Presbyterian Church raised $300 and V.M.I. Cadets contributed $421. One of the most impressive charitable efforts was that of "four little girls [who] set out with the determination to raise money enough to buy a barrel of flour for the hospital in Staunton. They came home in the evening with nearly $100.00 contributed by the citizens. And besides this they procured separate subscriptions of four hams, bread, &c."

Once a week a wagon loaded with edibles and other supplies left Lexington for the hospital in Staunton.[22]

War and its adjuncts were not the only concerns of the town in late 1862. City services were deficient, according to the editor of the *Gazette,* who admonished the city council to repair the streets before the advent of winter. he wanted "gutters opened, side walks repaired, loose stones removed and gulleys and depressions filled in."

That was not all. Police and fire protection was inadequate. Specifically, he noted, the Town Sergeant was unable to patrol the village satisfactorily at night. And when Mr. Shirley's smoke house was all but consumed by flames, the journalist noted, somewhat sarcastically, he expected "to hear the old engine rattling down the street; but the Fire Company must have concluded that a little smoke-house was utterly too insignificant an object, upon which to waste their energies and water."[23]

John S. Wise, a son of General Henry Wise, was a product of Tidewater Virginia. When he arrived in Lexington to enter the Virginia Military Institute in 1862, the observant teenager noted the influence of the Scotch-Irish settlers on the face of the village. He wrote:

> Their impress was upon everything in the place. The blue limestone streets looked hard. The red brick houses, with severe stone trimmings and plain white pillars and finishings, were still and formal. The grim portals of the Presbyterian church looked cold as a dog's nose. The cedar hedges in the yards, trimmed hard and close along straight brick pathways, were as unsentimental as mathematics. The dress of the citizens, male and female, was of single-breasted simplicity; and the hair of those pretty Presbyterian girls was among the smoothest and the flattest things I ever saw. . . . Looking eastward from the front of the tavern where the stage-coach deposited us, the barracks, mess-hall, professors' houses, parade ground, and limits of the Virginia Military Institute were in view upon a hill about half a mile distant.[24]

To the editor of the *Gazette,* life in Lexington was growing grim and deadly. He wrote, "The camp fever drives its victims home, not merely to die himself, but also to carry deadly disease into his own household. The old and young[,] male and female become truly the prey of this insatiable war." He added this opinion on the length of obituaries: "We ask them to be brief records of what their friends might have been. . . . Brief obituaries are always read — *long ones* very seldom."[25]

Other than the V.M.I., only three schools were reported operating in the county in the fall of 1862. Washington College resumed classes on September 16, Cornelius C. Baldwin's Forks School on September 22, and the Ann Smith Academy on October 1.[26]

In the world of commerce, Colonel Robert H. Brown of the Rockbridge Woolen Factory was said to be selling materials to consumers well below the prices prevailing in Lynchburg and Richmond. Later Brown offered to exchange cloth for lard and soap.

According to the *Gazette*, the Lexington Chemical Works "are now making blacking equal to Mason's celebrated blacking, and writing fluid of the first quality." A new tannery, appropriately named "Stone-wall Manufacturing Company," was erected at Buffalo Mills, while dog skins were being tanned and made into small leather items.

Inhabitants also were urged to remove the soil under smoke houses in order to extract salt petre. Furthermore, the citizens were urged to send wagons to the Kanawha salt works to bring back as much salt as possible while the Confederates held the area.[27]

Although Federal forces were not in Rockbridge County at the time, there was a Yankee problem, according to a letter in the *Gazette*. Allegedly, there were Yankee deserters employed by Joseph R. Anderson & Company in the iron works at Arnold's Valley. These defectors, the writer went on, were "tampering with negroes and committing other 'offences' in the area." Five of the deserters were already in jail in Lexington, he added, and demanded that they be expelled from the county.[28]

William Nelson Pendleton, the preacher-artillerist, also had a request to make of the folks back home. Writing from his post with the Army of Northern Virginia, the brigadier general asked for blankets, curtains, carpets, anything that could be used to cover soldiers during the forthcoming winter, as well as shoes and woolen socks. In response to the plea, collection points were set up at "A. Alexander at Jordan's Point; William White & J. M. [T.] McCrum, Lexington; J. W. Gilkeson, Brownsburg; Captain W. C. Gilmore, Kerr's Creek, & J. P. Lackey, Poague's Store." The appeal elicited the customary generosity, "175 blankets & other coverings, 75 pr socks, 50 pair of shoes (makings), leather for 50 more pair. . . $750 cash."

Upon receipt of the items, commanders of the various companies expressed their gratitude through a letter to the *Gazette*. Captain James J. White of the Liberty Hall Volunteers assured the donors that all the Rockbridge men were fully supplied. In addition, 67 blankets and coverings had been given soldiers from northwestern Virginia. Another plea, this one from Captain Frank W. Henderson of the Subsistence Department, urged people to sell him provisions for the army.[29]

After the hospital at Washington College met resistance, the Confederate government rented the buildings at the Lexington Fairgrounds for $300 a year to serve as a replacement facility. This arrangement inspired a warning from the weekly journal. "The establishment of a hospital here," wrote the editor, "increases a hundred fold the probabilities that smallpox will be brought to the county before many weeks. It is therefore a matter of the highest importance that families give attention at once to vaccination."[30]

The sanguinary repulse of Union assaults at Fredericksburg on December 13, 1862 took a heavy toll of life in Rockbridge units, especially the First Rockbridge Artillery. Through a letter from General Jackson, Margaret Preston learned that a cousin, fighting for the Union, had been killed in that affair. Writing in her diary of Christmas, she noted: "The sadness of the household forbids any recognition of Christmas; we are scattered to our separate rooms to mourn over the contrast [with peacetime Yules] and the Library is in darkness."

On December 31, Mrs. Preston penned: "Servants away all day on their holiday, and I have been doing much of their work. . . . God grant us a happier year in the one to come than the last has been."[31]

CHAPTER IV

"The Loss Was Irreparable"

The campaign by the *Gazette* to persuade the County Court to provide financial relief for soldiers' families as well as the indigent of Rockbridge brought results at the start of the third year of the war.

The court appropriated $15,000 for the purpose, the money to be doled out in this fashion: "Each woman to receive $1.25 per week, each daughter 12 years 75¢, each child under 12 50¢ per week." With the depreciation of Confederate currency, and with potatoes selling for five dollars a bushel, this would help a little but not much. The same rates applied to families of slaves drafted to work on fortifications.[1]

In another action, the court ordered that $10,000 worth of notes be issued for currency in amounts of 10 cents, 15 cents, 25 cents and one dollar and fifty cents. Each piece of currency was required to bear the signatures of the Justices of the Peace and the Clerk of the Court.[2]

Furthermore, the court ordered Sheriff David J. Whipple and his deputies to enroll all male slaves between 18 and 45 for work on the fortifications around Richmond. Owners were required to take an oath affirming the incapacity of any slave excused from such duty. On January 14 Whipple reported that 608 slaves were liable to be drafted. The slaves who were selected were ordered to report to Lexington, Brownsburg and Ore Bank the following week ready to proceed to Richmond. Later a report circulated that the slaves were improperly fed and housed. When a representative of the county investigated, however, he found the report groundless.[3]

In a municipal election, George W. Adams was chosen mayor of Lexington. the new city council consisted of Robert I. White, William G. White, James M. Pettigrew, David L. Hopkins, John G. Pole and George A. Baker. Commenting on the results of the election, the editor of the *Gazette* wrote: "We hope the new broom will sweep clean and that we shall soon have all our streets and roads 'put to rights.'

We would suggest the propriety of employing some of the refugee hands now here to work the roads in the vicinity of the town."[4]

When not overwrought by the condition of the roadways, the editor appealed for materials to aid and comfort patients at the Fairview Hospital on the Fairgrounds. Although conceding that "valuable articles of clothing, Bedding and Diet" had already been contributed by various aid societies, the newspaperman emphasized:

> The articles most needed are *Pillows, Bed-comforts, Towels, Combs* and *Soap*. The soldiers regard *corn bread* as a decided luxury. If some of our liberal farmers would therefore send in a few bushels of good meal occasionally, they would confer a real favor. *Buttermilk* too, is a great treat to convalescent patients. For the convenience of those making donations, Mr. J. W. Barclay has kindly offered to take charge of such articles as may be sent to his store-room. . . . The ladies of the town will attend to having all articles promptly distributed.

At a public meeting in the Lexington Presbyterian Church, ways were discussed to raise funds for the purchase of Bibles, testaments and religious tracts for servicemen. Reverend William M. McElwee, president of the Rockbridge Bible Society, appointed a committee to solicit funds for the endeavor and the local ministers were encouraged to ask their congregations to contribute to the effort. Chaplain Robert I. Taylor announced that the Fairview patients were extremely pleased to receive the Bibles. He added that there were 150-160 men in the wards, most of whom were convalescents.[5]

A few weeks later the dreaded smallpox surfaced at the hospital. There were three cases, and also three deaths from other diseases. The hospital was quarantined immediately, and the women of the vicinity who had been nursing the sick and wounded were barred from the grounds. In addition the town council adopted an ordinance forbidding the officers, inmates and employees of the hospital from entering the town.[6]

Inflation was tightening its grip on Rockbridge folks in no uncertain manner. Margaret Preston lamented her having to spend twenty-five cents apiece for seed potatoes and twenty dollars for a bushel of onions. Firewood was ten dollars a cord if delivered at Jordan's Point or twelve dollars if delivered in Lexington. Unbleached cotton was now two dollars a yard, handkerchiefs that formerly sold for ten cents were now going for two dollars and fifty cents. Thin cotton stockings were four dollars a pair and children's gingham aprons were six dollars.[7]

All the while a steady stream of refugees was arriving in Lexington. Margaret Preston revealed that she had heard of four new families in the village, "one is a mother with eight children." The population in the county, including refugees, was estimated at more than 17,000 persons.[8]

Among the many pieces of real estate up for sale at the time was a house and lot belonging to Stonewall Jackson. It was being offered by his agent, Colonel J. T. L. Preston, "Old Jack's" peace-time colleague on the V.M.I. faculty.[9]

On February 27, Samuel F. Atwill, a Cadet at the Institute, noted in his diary:

> To-day was set apart by the Pres of the Confederate States as a day of humiliation, fasting — and prayer; and of course we were compelled to attend church (the Presbyterian as usual), the day being very muddy and sloppy there was an unusual display of *legs* by the young ladies of the Seminary [Ann Smith Academy].[10]

During the March meeting of the County Court a committee was appointed "for the purpose of making a roll, showing the names of all officers, non-commissioned officers and privates, who have been, are now, or may be in the military service of the Confederate States, from the County of Rockbridge." The action was taken to preserve the names of deceased soldiers to be placed on a proposed monument. At the same meeting the court appointed Robert I. White as the county agent to visit the South and purchase cotton yarn and cloth for the use of Rockbridge housewives.[11]

For several weeks in March Cobb's Georgia Legion and other units of Wade Hampton's cavalry camped near Brownsburg, where they were treated with uncommon hospitality. Acknowledging the kindnesses, Captain G. I. Wright, spokesman for the regiment, announced the resolution adopted by the men: "1st. That we do hereby tender, to the citizens and ladies particularly of Brownsburg and vicinity, our heartfelt thanks for the kind offices they have generously performed in ministering to the wants of our sick, and the uniform courtesy which has been extended to the members of the Regiment."[12]

Rising temperatures in Rockbridge were clear indications that the armies would soon be launching their spring campaigns. On April 2, Cadet Atwill observed:

> Old "Sol" came forth this morning in all his glory — the snow is thawing very rapidly — sloppy underfoot — Twelve hundred cavalrymen passed here to-day on their way to North Carolina to recruit their horses and men — They seemed to be in fine spirits — some two hundred were dismounted — they lost their horses in the past six weeks.[13]

In response to a proclamation by President Jefferson Davis concerning food, especially meat, to feed the armies, a meeting was held at the courthouse on April 13, chaired by Judge John W. Brockenbrough. In the course of the confab, committees were formed in each magisterial district to visit every family in the county and determine the amount of surplus food that could be released to sustain the soldiers. This was done, in part, to prevent the government from impressing foodstuffs. The editor of the *Gazette* informed the gathering that he had received a letter in which an officer in the Army of Northern Virginia stated; "I have never seen this army so devoted to the cause as now."[14]

On April 22, Governor John Letcher announced his candidacy for the Confederate Congress. Six days later, accompanied by a coterie of army officers, he visited his home town, where he reviewed the V.M.I. Cadets. After the parade, Letcher inspected the Cadets' quarters and was introduced to the future officers. Though carrying Rockbridge County at the polls, Letcher was defeated by incumbent John B. Baldwin of Augusta County.[15]

With the arrival of spring, the Army of the Potomac, under General Joseph Hooker, crossed the Rappahannock and clashed with the Army of Northern Virginia, under Robert E. Lee, at the battle of Chancellorsville. From the thunderous artillery sounds that were clearly audible in Rockbridge County, inhabitants knew that a mammoth battle was in progress. By May 5 some details of the engagement and the names of some casualties were known to Margaret Preston, who wrote:

> Gen Frank Paxton is killed; Jackson and A. P. Hill wounded. Of the mothers in this town, almost all of them have sons in this battle; not one lays her head on her pillow this night, sure that her sons are not slain. The suspense must be awful. Mrs. Estill has four sons there; Mrs. Moore two; Mrs. Graham three, and so on. Yet not a word of special news, except that a copy of General Lee's telegram came, saying, a decided victory, but at great cost. God pity the tortured hearts that will pant through this night! And the agony of the poor wife who has heard that her husband is really killed![16]

Initial reports on battle casualties gave only a faint hint of the enormous harvest of death. News that Jackson's left arm, wounded accidentally by a volley from his own troops, had been amputated, was sufficient reason for grief. But even darker tidings followed. Captain Greenlee Davidson, commanding the Letcher Artillery, had been killed. The body of the young officer arrived in Lexington on May 9 and, after lying in state in the Society of Cadets Hall at the Institute, it was buried the same afternoon. According to Cadet Atwill: "The Corps attended his funeral under arms — had to march in the middle of the street in mud up to our ankles."[17]

So overwhelming was the anxiety of the people of Lexington that it was impossible for them to contemplate anything except war. On Sunday, May 10, Mrs. Preston noted: "This afternoon Dr. White attempted to hold service; but just as he was beginning, the mail arrived, and so great was the excitement, and so intense the desire for news, that he was obliged to dismiss the congregation. We only hear of one more death among Lexington boys. . . . Several wounded."

Mrs. Preston was in the progress of writing to Jackson and suggesting that he return to Lexington and recuperate in her home when she receive'd the shocking news that the general had died at Guinea Station on May 10. "My heard overflows with sorrow," she wrote. "The grief in this community is intense, everybody is in tears. . . . How fearful the loss to the Confederacy."[18]

Funeral services for General Frank Paxton, killed at Chancellorsville on May 3, were conducted in Lexington on May 12. The V.M.I. Cadets escorted the body to the Paxton home, three and one-half miles out of town, and back to the town cemetery where it was buried with full military honors.[19]

Cadet Atwill described the arrival and burial of Jackson's body as follows:

> Thursday May 14th — Gen Jackson's body arrived by the boat at 1 o'clock — was escorted to the barracks by the Corps and placed in his old Lec[ture] room — which room is draped in mourning for the period of six months — he is in a fine metallic coffin — the *first flag* made in the south of the new design covers his coffin — on the flag wreaths of evergreens and flowers — it is the request of his wife that he shall be buried tomorrow — half hour guns have been firing from sunside [sunrise] fired from his old battery. Colonel Smith also directed the officers and cadets to wear mourning bands for 30 days.[20]

Cadet John S. Wise recorded his observations of the memorable occasion:

> It was a bitter, bitter day of mourning for all of us when the corps marched down to the canal terminus to meet all that was mortal of Stonewall Jackson. . . . With reversed arms and muffled drums we bore him back to the Institute, and placed him in the section-room in which he had taught. . . . The number of people who came to view him for the last time was immense: men and women wept over his bier as if his death was a personal affliction; then I saw that the Presbyterian could weep like other folks. The flowers piled around the coffin hid it and its form from view. I shall ever count it a great privilege that I was one of the guard who, through the silence of the night, and when the crowds had departed, stood watch and ward alone with the remains of the great "Stonewall."
>
> Next day, we buried him with pomp of woe, the cadets his escort of honor; with minute-guns, and tolling bells, and most impressive circumstance, we bore him to his rest.[21]

Cadet Atwill recorded the lachrymose circumstances as they came under his observation:

> The procession formed in front of the Sally port at half

past ten, commenced to move at 11 — corps in front of caisson on which he was borne — there a company of Cavalry — after that a company composed of all the wounded and all were. . . members of the old Stonewall Brigade.[22]

The *Gazette* account of the order of procession was as follows:

The body reached Lexington by the Packet boat . . . accompanied by his personal staff, Maj. A[lexander] S. Pendleton, Surgeon H. [Hunter] McGuire, Lieut. [Joseph] Morrison and Lieut. [James Power] Smith, by His Excellency Gov. Letcher, and a delegation of the citizens of Lynchburg. . . . The Funeral escort was commanded by Major S. [cott] Ship[p], Commandant of Cadets, a former pupil of Gen. Jackson and a gallant officer who had served with him in his Valley campaign, as Major of the 21st Va. Reg't. The escort was composed as follows:

1. Cadet Battalion. 2. Battery of Artillery of 4 pieces, the same battery he had for ten years commanded as Instructor of Artillery and which also had served with him at 1st Manassas [1st Rockbridge Artillery] in Stonewall Brigade. 3. A company of the original Stonewall Brigade, composed of members of different companies of the Brigade, and commanded by Capt. A[lexander] M. Hamilton bearing the flag of the "Liberty Hall Volunteers." 4. A company of convalescent officers and soldiers of the army. 5. A squadron of cavalry was all that was needed to complete the escort prescribed by the Army Regulations. This squadron opportunely made its appearance before the procession moved from the church. The Squadron was a part of Sweeney's battalion of Jenkins' command, and many of its members were from the General's native Northwestern Virginia. 6. The Clergy. 7. The body enveloped in the Confederate Flag, and covered with flowers, was borne on a *caisson* of the Cadet Battery, draped in mourning. The pall bearers were as follows: Wm. White, Prof. J. L. Campbell, representing the elders of the Lex. Pres. Ch.; Wm. C. Lewis, Col. S. McD. Reid, County Magistrates; Prof. J. J. White, [Prof.] C. J. Harris, Washington College; S. McD. Moore, John W. Fuller, Franklin Society; George W. Adams, Robert I. White, Town Council; Judge J. W. Brockenbro', Joseph G. Steele, Confederate District Court; Dr. H. H. McGuire, Capt. F. W. Henderson, C. S. Army; Rev. W. McElwee, John Hamilton, Bible Society of Rockbridge. 8. The Family and Personal Staff of the deceased. 9. The Governor of Va., Confederate States Senator Henry of Tenn. The Sergeant-at-Arms of C.S. Senate, and Member of the Richmond City Council. 10. Faculty and Of-

ficers of Va. Mil. Institute. 11. Elders and Deacons of Lexington Presbyterian Church of which Gen. Jackson was a Deacon. 12. Professors and Students of Washington College. 13. Franklin Society. 14. Citizens.[23]

Margaret Preston, the general's sister-in-law by his first wife, added:

> The exercises were very appropriate, a touching voluntary was sung with subdued, sobbing voices; a prayer from Dr. Ramsey of most melting tenderness; very true and discriminating remarks from Dr. White, and a beautiful prayer from W.[illiam] F. J.[unkin]. . . . The grave too was heaped with flowers. And now it is all over, and the hero is left "alone in his glory." Not many better men have lived and died. His body-servant [Jim Lewis] said to me, "I never knew a piouser gentleman." Sincerer mourning was never manifested for any one, I do think. . . . The dear little child [daughter Julia, born November 23, 1862] is so like her father; she is a sweet thing and will be a blessing, I trust, to the heartwrung mother.[24]

"The burial of Stonewall Jackson made a deep impression upon the corps of cadets," recalled John Wise. "It had been our custom, when things seemed to be going amiss in the army, to say, 'Wait until Old Jack gets there; he will straighten matters out.' We felt that the loss was irreparable."

Following the funeral, Mrs. Jackson and her daughter spent several weeks with the Prestons before returning to the home of her parents in North Carolina. According to Elizabeth Randolph Preston, "the privilege of sheltering the heart-broken wife in the first bewilderment of distress was one beyond valuation to the . . . household."[25]

Before returning to Richmond, Governor Letcher was asked to chair a meeting at the court house, at which it was proposed to erect an equestrian statue of Jackson atop a monument listing the names of deceased Confederates from Rockbridge County. Committees were formed to canvas the area for donations to the project.[26]

While the South grieved over the loss of Stonewall Jackson, its collective concern turned to more immediate matters, like the decline of the Confederate dollar's purchasing power. Margaret Preston lamented the $75 she was forced to spend for a plain crepe bonnet, and $180 for a bombazine dress. "Bought . . . a ninepiece calico dress. . . for which I gave $30," she related. "Unbleached, very coarse cottons are now $2.25 per yard. . . . Mr. P. paid for some days' work of a white man, a short while ago, at $8 per diem."

The *Gazette* was hard-pressed to obtain paper in order to continue its weekly publication. In desperation, the editor offered five cents a pound for rags of any color. Comparing peace-time conditions with those of the present, the journal went on:

> Three years ago he [an anonymous individual] gave us a cord of wood for one year's subscription — now we will take less than half a cord. He used to pay his subscription with twelve pounds of butter, now he can do the same with four pounds. It formerly required four bushels of corn, now we will send him the paper for one bushel. While we would have demanded sixteen chickens for our paper three years ago, we will now be content with four, half grown specimens of the feathered tribe, though not sufficient for a comfortable breakfast for ourselves and the employees of our office. If you wish to pay in hay or flour, we will take one-fourth of what you formerly gave us.
>
> You say the paper has been reduced in size. So it has; but still our material costs us more than three times as much as it formerly did. Are we not right then in advancing our rates to what they now are? And may we not be justifiable in another advance within the next three months?[27]

With the Confederate victory at Winchester (June 15) and subsequent advance through Maryland and into Pennsylvania, Mrs. Preston wrote:

> Hear that Lee's army is invading my native State. Well! Virginia has endured it for more than two years! So I must not think it hard that another state whose troops have been helping to ravage her all this time, should take its turn. I trust this army will not be guilty of the outrages which have everywhere characterized the Federal armies in Virginia. . . . Our town is so full of refugees, people who have fled from their homes, that I scarcely know anybody I meet.[28]

The crucial battle of Gettysburg cast another cloud of sadness over the homes of Rockbridge County. The casualties in the county's two artillery batteries were reported initially in the *Gazette.*

A member of the Liberty Hall Volunteers wrote:

> Our company went into the engagement with 21 rank and file, of whom there are now only three in camp, the rest have been all killed, wounded or captured. . . . Lt. Jones is the only one who went near the works [of the enemy] and returned.

The Rockbridge Grays lost two men killed. Captain McCorkle and the rest, save two men, were captured.

Uncertainty enveloped the household of the Preston family.

Margaret learned early that a cousin, a captain, had been killed, but it was a year more before she learned positively that the captain's younger brother, a lieutenant, had also met his end in the three-day struggle in Pennsylvania.[29]

With commissioned personnel falling in ever increasing numbers, Washington College authorities tried unsuccessfully to establish a training school for officers on its campus in the wake of the Gettysburg defeat. However, the faculty set up a course of military training and appointed a committee "to procure from the Confederate States authorities, a suitable supply of Arms for the use of the students in the Department of Military Tactics." By instituting such a program, the faculty hoped to keep students on campus instead of enlisting in the armed forces.[30]

Due to the sinking fortunes of the Army of Northern Virginia, absenteeism shot upward. In an effort to check the trend, Lieutenant Henry A. Wise led a detachment of V.M.I. Cadets, commanded by Captain Robert D. Lilley, the county Enrolling Officer, in a search for the deserters. They scoured the mountain areas diligently, but came away empty-handed. Lieutenant Colonel James K. Edmondson, the Provost Marshal of Lexington, also issued a strong appeal to the AWOL men, but this, too, was fruitless.[31]

A late-August raid into western Virginia by Union cavalry under William W. Averell sent reverberations all the way to Lexington. When the Federal advance reached Warm Springs in Bath County on August 30, Colonel William L. Jackson's Brigade of southern horsemen fell back to Warm Springs Mountain.

Jackson was opposed by superior numbers, a fact he readily understood. He sent an urgent appeal for help to Edmondson and to Francis H. Smith at the V.M.I. Edmondson responded by sending the Home Guards and a detachment from his provost guard. Smith, in turn, sent two companies of new Cadets with a four-gun battery under Captain Wilfred E. Cutshaw.

The two commands left Lexington at noon on August 25 and marched 11 miles to Rockbridge Baths, where they went into bivouac. The following day Edmondson and the cavalry joined Jackson. Smith, with the Home Guards and Cadets, marched to Goshen and later to Panther Gap. When Jackson asked Smith to help in an attack on Averell, the V.M.I. superintendent declined, explaining that the Board of Visitors had restricted the service of the Cadets to defensive action in Rockbridge County. Learning that Averell was withdrawing northward, Smith led the Cadets and Home Guards to Rockbridge Alum Springs, where they spent the night. The command returned to Lexington on August 27, at the close of a 50-mile exercise.

Cadet Beverly Stanard reported that the new Cadets were "broken down and sorefooted, and quite mad that they were not permitted to go on and engage the enemy." Later Governor Letcher lifted the restriction on the use of the Cadets in military action.[32]

The Home Guard companies that had served under Edmondson and Smith were officered as follows: Company A, Captain Thomas H. Williamson, 1st Lieutenant James J. White, and 2nd Lieutenant James T. McCrum. Company B, Captain William Gilham, 1st Lieutenant Jacob M. Ruff, and 2nd Lieutenant George W. Adams. Each company numbered more than 50 members.

As a result of the phantom encounter with Averell, the *Gazette* missed its weekly publication date. The editor explained:

> Although our paper is dated for Wednesday, we failed to issue it on that day, in consequence of all hands connected with the office having turned out with the Home Guard to defend the County.
>
> Editors and printers all took up arms without exception; and so shall always be ready to do, when the homes of our people are threatened.[33]

In quick order, other districts of Rockbridge organized Home Guard units. Captain Alexander D. Campbell drew his men from the Fourth District — Buffalo and Colliers creeks — with Bolivar Leech and John Holden as lieutenants. In the Second District, Buffalo Forge and Fancy Hill — Captain William F. Junkin organized his 100-man unit with Lieutenants James G. Updike, Marius M. Hartsook and Charles W. Michie. At Brownsburg, Captain Zachariah J. Walker enlisted his company of mounted infantry, aided by Lieutenants James J. McBride and Daniel Brown. In the Seventh District, Captain William A. Donald and Lieutenant John M. Templeton raised a company, as did Captain Alexander M. Glasgow and Lieutenant Samuel Gibson. At the Collierstown Nitre Works, Captain James L. Leitch formed a unit of 56 men with the assistance of James E. A. Gibbs and John A. Leech. These companies were banded into a regiment in October. James W. Massie was elected colonel, William F. Junkin the lieutenant colonel and Daniel Brown the major.[34]

With the decline of civilian manpower as a result of military requirements, ever stronger demands arose for the services of slaves and freed men. Some of the emanicipated blacks worked in the nitre caves of Bath, Highland and Pendleton counties, so that white conscripts detailed for that work could return to the army. Captain James F. Jones, of the Nitre and Mining Bureau, wrote from his office in Staunton: "You can assure the negroes that they will be well fed, kindly treated, and paid sixty cents per day."

Captain Thomas H. Tutwiler also detailed ten free men to work for the Quartermaster Department in Lexington. The slaves were assigned to Richmond, where they worked on the fortifications around the city.[35]

The reorganization of the Home Guards was completed none too soon, as Averell launched another raid into western Virginia on November 1. The extent of the raid was not known in Lexington until November 5, when the Cadets and Home Guards were called into action. Colonel William Jackson reported the enemy was again near Warm Springs. Couriers raced through Rockbridge sounding the alarm. The Home Guards assembled the next morning in Lexington to draw arms and ammunition. A company of students from Washington College under Captain Alexander L. Nelson and another from Kerr's Creek under Captain John McKemy also mustered with the old men and young boys. Colonel Scott Shipp commanded the Cadet battalion and Lieutenant Thomas H. Smith the Cadet battery.

"This was the force," noted the *Gazette,* "and until we saw it together we had no idea there were so many men in the county."

Mrs. Preston commented, "The whole town is in commotion; no man left in it; even those over sixty-five have gone."

A third report, by H. McC. Davidson, added, "Every body big and little, old and young turned out, there was no holding back."[36]

The two commands marched to Rockbridge Baths, where they camped for the night. The next morning the Home Guards and Cadets trekked to California Furnace. On November 8, the march resumed, the column winding over the mountain to near Lucy Salina Furnace. Here General John D. Imboden, commanding the Valley District, learned of the Confederate reverse at Droop Mountain. He ordered Massie and Shipp to move their units to Covington. The two commands, now under Francis H. Smith, who had joined the force en route, marched at daybreak on November 9. Proceeding westward, they halted that night at Clifton Forge. The two commands joined Imboden at Covington the next day.

The following day Imboden placed his combined forces on a hill a mile east of town. After a skirmish between Union and Confederate cavalry, and an exchange of artillery fire, the Federal general withdrew from the field. Learning that Averell was in full retreat, Imboden ordered the Home Guards and Cadets to return to Rockbridge.

Before dismissing the troops, Imboden ordered Massie to call them together for a word of commendation. The general thanked them for their prompt response to the raid and added: "If the Valley counties will but act the part that old Rockbridge has done on his occasion, there will be no need for reinforcements from Gen. Lee to repel raids against the Valley." The speech evoked enthusiastic applause by the old men and boys. In his official report of the action, Imboden stated:

> I beg leave to add my testimony to the admirable spirit displayed by the people of Rockbridge in coming to my assistance. At 7 p.m. on Thursday, the 6th, the news reached Lexington of this raid. By 7 p.m. the next day 800 men were 12 miles on their march to support me. My thanks are especially due Col. J. W. Massie, commanding the home guards, and General F. H. Smith, commanding cadets, for their energy and zeal they manifested and the skill which which they moved their commands so rapidly through the mountains.

Because of inexperience, many of the Home Guards had embarked on their expedition without adequate clothing and blankets. As a result, much suffering was endured in cold and snowy conditions.[37]

Meanwhile, inflation continued its mad spiral upward. Mrs. Preston noted that wood was selling for $30 a cord, flour $50 a barrel, butter $3.50 a pound, and brown sugar $3 a pound. Silk for a wedding dress was priced between $500 and $600. The Lexington Tannery, she reported, had generously donated enough leather for 100 pairs of shoes for families of servicemen. The County Court pitched in by appropriating another $30,000 for indigent families of Confederate soldiers. Shortly, the amount for each dependent was doubled. The court also appointed agents to purchase cotton cloth and yarn and other articles to be sold to the needy. Agents were also appointed in each magisterial district in the county to receive and distribute the items.[38]

Another segment of society hit hard by inflation was the medical profession. Doctors adopted resolutions stating they would treat patients at the prewar rate if the people compensated them with foodstuffs or other articles that could be used at the prewar rate. Patients also were required to furnish the goods within six months. Those who chose to pay cash for the doctors' services or provide items at current prices would be charged prevailing prices. The only exceptions to these practices were the families of soldiers or impoverished folks. The resolutions were signed by 24 physicians practicing in the county.[39]

Patrols to maintain law and order, and more precisely to prevent night-time movements of slaves between plantations around Lexington, which had become a fairly common occurrence since the start of the war, were extended to outlying sections of the county.[40]

December brought an even more serious raid by Averell's Federal cavalry. Informed that Union horsemen were threatening Covington, Smith and Massie ordered out their commands on December 14. The troops filed out of Lexington at dawn the next morning and encamped at Rockbridge Baths that night.

On orders from Robert E. Lee, General Fitzhugh Lee led his cavalry division to the site of anticipated combat, where it was augmented by a small infantry command under General Jubal A. Early, now in command of the Valley District.

The V.M.I. Cadets reached Bratton's Run on December 16. They found the stream unfordable and bivouacked on the banks. The mounted Home Guards halted at Goshen. When Early learned that the enemy had already passed Salem and was headed toward Lexington, he ordered Massie to return to the Rockbridge community and defend it. At the same time Massie was directed to send scouts toward Buchanan, burn the bridge over the James River if necessary, as well as the bridge over Buffalo Creek. All the while, Imboden was en route to Lexington via Brownsburg. Massie, Imboden and Lee arrived at Lexington on December 18.

Margaret Preston described what happened next:

> At 11 o'clock, Imboden's cavalry and artillery passed through. It is the first time I have seen an army. Poor Fellows! With their broken down horses, muddy up to the eyes, and their muddy wallets and blankets, they looked like an army of tatterdemalions; the horses looked starved. Then came the Home Guard, drenched and muddy, as if they had seen hard service, though they had only been out four days; but such weather! It rained terribly, the rain part of the time freezing as it fell; and they were out in it all: stood round their fires all night, or lay down in puddles of water. At 3 p.m. Gen Fitzhugh Lee's cavalry (2700) passed through. Their horses were in better condition. All the men in both divisions looked in fine spirits, and cheered vociferously as the ladies waved scarves and handkerchiefs on their passing. People brought out waiters of eatables for the poor tired men. . . .All went to Collierstown for the night. . . . The Cadets were water bound, some of them waded to their waists in water, building bridges for the artillery. Mr. P. [Preston] saw one marching along in his *naked* feet. This is "glorious war!"

The cadets, under the leadership of Colonel Shipp, got the guns and wagons across the creek on December 17 and pushed on to Cold Sulphur Springs before camping for the night. Many lost their shoes in the ankle-deep mud.

The young soldiers returned to Lexington the next day, by order of General Smith, only to be ordered out again shortly thereafter. This time they marched to Wilson's Springs.[41]

While Fitz Lee was en route to Clifton Forge with his division and Imboden's Brigade, word was received that Averell was advancing on Buchanan. Lee reacted immediately. Diverting his command from

Averell's escape route, he led it to Buchanan, toward which the Cadets were marching at the same time.

The Cadets had left Lexington on December 20, an extremely cold day, and proceeded to Buchanan over frozen roads. Despite the heroic performances of all the Confederate commands, Averell eluded pursuit and returned to his base with minimal loss.

Fitz Lee was much impressed by the efforts of the V.M.I. contingent and expressed his thanks to Smith and the Cadets for their fine performance; also to William A. Mann, former sheriff of Rockbridge County, for his services during the emergency.[42]

Except for soldiers home on leave, Christmas provided scant holiday merriment in Rockbridge County. Mrs. Preston's preparations for the Yuletide contrasted sharply with prewar days, consisting of "crullers with molasses, and mince pies without sugar or fruit or spirit." The V.M.I. menu offered nothing better. According to Cadet Stanard, there was "cold loaf bread, . . . warm *cornbread*. . . Cold beef. For *dessert,* molasses, water &tc."

The Lexington restaurants fare was little better, but exceedingly more costly than in the mess hall. In an effort to forget their gastronomic problems, many Cadets and youngsters of the town spent many year-end hours skating and playing ice hockey on the frozen North River.[43]

Despite their own skimpy supplies of food and goods, people of Rockbridge gladly shared what they had with the soldiers. Their gestures were not unappreciated. James D. Davidson, the Commissary in Lexington, received a letter from General Imboden, thanking him for the provisions, gloves and other articles of clothing. "We shall never forget the hospitality of good old Rockbridge," wrote the cavalryman. "Every day I hear some misty looking veteran say, 'Well boys, I wonder when we will be marched toward Rockbridge again?' And often you will hear in camp, "Hurrah for the Rockbridge Home Guards.'"

Colonel William L. Jackson, nicknamed "Mudwall" in contradistinction to his second cousin, "Stonewall," expressed gratitude to the ladies of Lexington, Newport, Brownsburg, Timber Ridge and other locales for the clothing sent to his command at Warm Springs. Two members of Company I, 31st Virginia Infantry from Northwest Virginia, also wrote to thank "John Mohler, John & Andrew Snider & Mrs. E. Snider of Rockbridge Co. for nursing them" after being wounded in battle.[44]

CHAPTER V

"Rockbridge Invaded"

The fourth year of the War for Southern Independence was ushered into Rockbridge County on a sinister note — small pox prostrated a number of slaves and free blacks. The farms that the victims occupied were instantly quarantined, however, preventing the spread of the dread disease and eliciting a collective sigh of relief from the county residents.[1]

In the midst of their austere lifestyle, Lexington Episcopalians permitted themselves a muffled chuckle when William Nelson Pendleton, their former rector, returned to his old church and preached a Sunday morning sermon. The 55-year-old artillerist delivered his message in his gray uniform because, it was announced, some miscreant had purloined his clerical robe from the vestry room.[2]

The first weeks of the new year were also marked by the return of the First Rockbridge Dragoons to the county. The unit was disbanded in order to recruit its horses until March 15. Later, General Thomas L. Rosser's Brigade of cavalry arrived for the same purpose, camping near the Falling Springs Presbyterian Church. Rockbridge was chosen as the site for equine rehabilitation because of the bumper crops of hay and corn that were available for the mounts.

The abrupt appearance of a host of young males was pure delight to the lovelorn lassies of the area. Parties, dances and unmitigated merrymaking abounded despite a shortage of viands and libations that customarily accompanied such events.

When not socializing, the great majority of troops reenlisted. The First Rockbridge Artillery pioneered in this move, and was followed in short order by the Second Rockbridge Artillery. Soon all the county companies had renewed their pledge to carry the war to its bitter conclusion.[3]

Of primary consequence to Rockbridge residents was the passage of a bill by the Confederate Congress providing for new currency and bonds. The bill was designed to halt inflation by reducing the amount of money in circulation. The effort boomeranged. Inflation skyrocketed as a result of the public's reluctance to accept the old money. On February 19, Margaret Preston wrote:

> Everything is in an excitement about the currency bill, which we heard of last night. On the 1st of April it is to sink to two-thirds its present value, so everybody is trying to get it off

their hands. I have ceased noting the prices of things; they are so incredible . . . $30 per gallon for sorghum molasses; calico $12 per yard; tallow candles, $6 per pound, unbleached cotton, $5 per yard.

A week later the diarist added:

Sister pays $20 for having a home-made cotton dress made up. Unbleached cottons are $8 per yard. People are trading as far as possible, instead of paying money. As for example, the shoemaker tells me that he won't make a pair of shoes for me unless I send him a load of wood; so before the shoes can be had, his wood is sent. Flour is selling at $250 per barrel.

As prices soared even higher — coffee to $16 a pound and tea to $40 a pound — the County Court was forced to double its allowances for the dependents of soldiers.[4]

To counter more effectively Union raids such as those conducted by Averell's cavalry in the past, Major Thomas H. Williamson of the V.M.I. was directed to select sites in the western area of the county and beyond that were suitable for fortifications. In due time he wrote that Colonel William L. Jackson's engineers were fortifying Millboro and McGraw's Gap on Warm Springs Mountain in Bath County. Concurrently, General Imboden's engineers were erecting fortifications about seven miles from Goshen on the Cowpasture River, as well as at Buffalo and Jenning's gaps in Augusta County. Williamson also opined that Strickler's Gap, between Rockbridge Alum Springs and Millboro, needed to be fortified, as did Kerr's Creek and Colliers Creek gaps. In the next sentence, the officer wondered where all the troops would come from to man these positions.[5]

As the verdant season returned to Rockbridge, Rosser's troopers made ready to quit the county. Before leaving, they left a memento with the Corps of Cadets at the V.M.I. It was a captured Federal flag.[6]

On May 1, the ranks of the Home Guards were stripped of all able-bodied men who were subsequently enlisted in the Reserve forces of the county. The units were officered as follows: Lexington Company — Captain Richard C. Turpin and Lieutenants Algernon S. Bacon and James M. Pettigrew; Natural Bridge Company — Captain James G. Updike and Lieutenants N. J. Miller and William B. F. Leech; Fairfield Company — Captain William A. Donald and Lieutenants John M. Templeton and William G. Hamilton; Rockbridge Baths Company — Captain David P. Curry and Lieutenants Robert Hutchenson and Robert Montgomery.

All of the 17-year-olds were assigned to a company commanded by Captain Charles W. Freeman, himself a teenager, as were his lieutenants, Duncan C. Lyle and Alexander Laird. Major James J. White was placed in command of the battalion.[7]

Before the resumption of open warfare in the area, Lexingtonians paused to honor the memory of a fallen townsman. On May 10, the anniversary of his death, Stonewall Jackson was remembered by his former friends and neighbors. The highlight of the cemetery ceremonies was the raising of a flag that had been sent to Lexington by Hugh Seddon, a Londoner with Confederate sympathies. According to the *Gazette,* "Ex-Governor Letcher addressed an immense crowd of ladies and gentlemen assembled to witness the hoisting of the flag."[8]

Infirmary camps for horses that had been established at Brownsburg, Lexington, Collierstown, Fancy Hill, Natural Bridge and Lickum Bridge, were dismantled in the spring of 1864. The horses and the cavalrymen who had tended them were returned to their regiments.[9]

Indications that the major armies of the North and South were launching their spring campaigns were clearly audible to Lexingtonians on May 4. On that date, the Army of the Potomac, now under the command of Ulysses S. Grant, clashed with the Army of Northern Virginia. The sounds of that mighty confrontation in the Wilderness were heard by Shenandoah Valley inhabitants scores of miles away.

Closer to home, Union General Franz Sigel advanced his command up the Valley toward Staunton. His move was countered by John C. Breckinridge, commanding Confederate forces in western Virginia. The one-time Kentucky lawyer, who had been elected vice-president of the United States at 35, dispatched a frantic call for reinforcements. The Rockbridge Reserves and the V.M.I. Cadets responded quickly, marching out of Lexington on May 11.

Many horses from the vicinity of Lexington were impressed to transport the Cadet battery and pull wagons that hauled supplies to supply the soldiers.

On Sunday, May 15 the two armies met at New Market. By late morning Sigel was in retreat, due chiefly to a heroic charge by the Cadets. Of the 247 young soldiers who participated in the attack, ten were killed, including S. F. Atwill and J. B. Stanard. The Reserves were not involved in the battle.[10]

Rockbridge County was well represented in the ranks of the valorous Cadets. Company B was led by its one-armed captain, Frank Preston, a Lexington native and veteran of the First Rockbridge Artillery. Robert L. Brockenbrough, Joseph R. Echols, Samuel H. Letcher, James W. McCorkle, John H. Shields, and John S. and R. Joseph White were from the county. Porter Johnson's family had refugeed in Brownsburg from western Virginia.[11]

In eastern Virginia, Robert E. Lee was trying to withstand the Federal onslaughts in the Wilderness and around Spotsylvania Court House. He needed help badly. Word was sent to Breckinridge in the Valley to succor the Army of Northern Virginian with all the infantry and artillery available. While the Cadets marched to Richmond, Breckinridge complied with Lee's order, leaving only Imboden's cavalry units to defend the Valley.[12]

Daily the casualty reports from the East grew longer and were clouded with much uncertainty. To clarify the situation, Captain John C. Middleton and a reporter from the *Gazette* were sent to the site of conflict to look after the dead and wounded from Rockbridge. Middleton made a thorough check and sent back lists of names among the staggering losses, especially the missing and presumed dead. Rockbridge readers were horrified to read that the Stonewall Brigade, famed in song and story, was reduced to 200 men.[13]

The sanguinary struggle between Lee and Grant around Fredericksburg not only exerted a colossal strain on the endurance of the men, it also sorely taxed the Commissary Department as far away as Lexington. Captain Henderson, the county commissary officer, forwarded 92,000 pounds of bacon he had collected through tithes, 5,700 pounds he had purchased and another 10,000 pounds that he had gathered from the rest of Rockbridge which he loaned to the government. By 1865, Henderson estimated, another 100,000 pounds would be available, exclusive of the amount consumed by the cavalry when the horsemen recruited in the county.[14]

When Grant learned that Lee had been reinforced by troops from the Valley, he envisioned an opportunity to capitalize on the weakened defenses west of the Blue Ridge. He ordered David O. Hunter, who had replaced Sigel as commander of the Valley District, to march with his 12,000 men to Staunton, Charlottesville and Lynchburg. Imboden, outnumbered three to one, was no match for Hunter. He summoned the Reserves from Rockingham, Augusta and Rockbridge counties, while General William E. (Grumble) Jones, commanding in Southwest Virginia, assembled all the available men from his sector and marched them to Staunton.

When Jones arrived at his destination, he discovered that Imboden had already organized the force into regiments and brigades, and that Confederate cavalrymen were skirmishing briskly with Union horsemen. The Rockbridge Reserves were placed in Colonel Kenton Harper's regiment, which consisted of the older reservists of the three counties.

In the battle of Piedmont on June 5, the outnumbered Rebels nearly won the day. The balance swung to the northern arms when Jones, trying to plug a gap in his lines, was killed. When Jones fell, what had been a tightly contested battle turned into a rout. The old men and young boys from Rockbridge who were not killed, wounded or captured retreated to

Rockfish Gap and eventually to Lynchburg to help defend that city.[15]

The bloody repulse at Piedmont and the continuing slaughter of troops north of Richmond inspired the Rockbridge County Court to action. It appointed a committee that was instructed to proceed to the battlefields and make certain that the hometown casualties received the best possible care and attention. The committee consisted of Eli S. Tutwiler, Samuel F. Chandler, Dr. James L. Watson, Samuel Montgomery, Dr. John Cooper, Daniel Brown and Dr. John Alexander. Alternates were John W. Haughtawort, James R. Saunders, O. C. R. Burks, Lafayette Sehorn, William Jordan, Dr. William F. Humphreys and John Wallace.

Another committee was organized to receive supplies and distribute them to the committee in the field. Colonel Samuel McD. Reid headed this group, assisted by Hugh Barclay, Alexander L. Nelson and Hugh S. Wilson.

The *Gazette* recommended that two men from each magisterial district be sent immediately to the battlefields to administer to the wounded, and arrange for their removal to hospitals. The weekly journal pointed out that a similar committee from Lynchburg had helped care for about 8,000 casualties. Dr. Samuel M. Dold, Haughtawort, Tutwiler and William C. Gilmore were soon en route to Richmond.[16]

While Rockbridge citizens were directing their attention eastward, a new menace was brewing in the West. A strong Union force comprising General George Crook's infantry division and Averell's cavalry division was closing in on Warm Springs and Covington.

All that stood between the invaders and Rockbridge were two cavalry brigades. One was commanded by Colonel "Mudwall" Jackson, the other by newly-promoted Brigadier General John McCausland. The son of an Irish immigrant, McCausland had graduated at the head of his class at V.M.I. in 1857 and later taught mathematics at the school and the University of Virginia.

Prospects for Confederate success were at low ebb when fortune smiled on McCausland. He captured a copy of instructions to Crook. Aware that Crook and Averell planned to join Hunter at Staunton and take part in the capture of Charlottesville and Lynchburg, McCausland devised a scheme of his own to thwart the Federal strategy. First, he rejected pleas by his regimental commanders to take on the enemy wherever they should meet. Second, he decided to employ Fabian tactics, avoiding a direct conflict, even as Fabius had checkmated Hannibal more than 2000 years earlier.

"John Tiger" McCausland forced Crook and Averell to use most of their men to flank him out of his stronghold at Panther Gap. Next, McCausland fell back to Goshen and retired slowly up the railroad toward

Staunton, camping in Bell's Valley. On June 6, he arrived at Buffalo Gap, where he learned of Grumble Jones' setback at Piedmont. Marching his troopers around Staunton, McCausland came to the Middlebrook road, down which he retreated to near Brownsburg, where he learned of the junction of Crook and Averell with Hunter.

The Confederates proceeded through Brownsburg, where their band presented a concert for the residents, and moved into Augusta County, where they took a position near Arbor Hill. Because Hunter was operating under discretionary orders, and could advance on Lynchburg either through Charlottesville or Lexington, McCausland had the option of either facing or following "Black Dave."[17]

As the antagonists maneuvered nearby, Rockbridge citizens were thrown into near panic. Federals were reported at Jordan's [California] Furnace with no Confederate troops to harass them. As a precautionary move, some books were removed from the V.M.I. library and stored at Washington College. Families hid their jewelry and silverware. Livestock, slaves and food were sent to the mountains for safekeeping.

Couriers circulated reports that Staunton was in flames and refugees, pouring into Lexington from the northern sector of the county, added to local frenzy with additional tales of destruction. At the V.M.I. officers shipped arms, munitions and equipment to Lynchburg by canal boats. The Cadets returned to the Institute by packet on June 9. The battery, wagons and impressed horses returned later by the overland route.[18]

Hunter's four columns, totalling about 18,000 effectives, left Staunton for Rockbridge County on June 10. As the main Union force took the road to Greenville, Crook and Averell followed the Brownsburg road, along which they skirmished with McCausland. Another Federal cavalry division, under General Alfred N. Duffie, headed toward Waynesboro, where it met with Jackson's horsemen. Concluding that the area was too strongly held, Duffie advanced to Mount Torry Furnace, which he put to the torch. From there he crossed the Blue Ridge at Tye River Gap, and destroyed a Confederate wagon train and the Arrington railroad station before moving on to Amherst Court House.[19]

While Duffie was wreaking havoc east of the Blue Ridge, McCausland was contending with Crook and Averell a mile outside of Staunton. "John Tiger" gave his adversaries all they could handle, forcing them to dismount and form a skirmish line before yielding ground. The process was repeated until the two forces neared Arbor Hill. There McCausland posted his men behind fences and resisted the Federals stubbornly for three hours before he was flanked from his position. Through Middlebrook and Newport to Brownsburg, McCausland disputed passage of superior northern forces. "During the day we repelled 5 charges & made two to keep them from closing upon our rear," reported one of Mc-

Causland's officers.

Averell turned off at Newport. He rode to the Walker's Creek Valley and Hay's Creek, attempting to get to the rear of McCausland. The hit-and-run fighting continued to Brownsburg, where Averell was engaged again. Once more the Rebels retreated, this time to the Fairfield-Lexington road. McCausland camped at the Cameron farm, about two miles from Lexington. Crook halted his division near Brownsburg, with pickets at Cedar Grove. Averell bivouacked on Hays Creek, with his headquarters at "Bellvue." Hunter, who was travelling with General Jeremiah C. Sullivan's infantry division, ordered the destruction of two flour mills at Greenville, one at Midway and one at Fairfield.[20]

Calvin Culton and a black companion were playing at the family home on Hays Creek when he heard the cry, "The Yankees are coming!" The youths raced to the house, where they were joined by a second Culton son. The three were handed hams from the smoke house with orders to conceal them from the despicable marauders. The youths rode into the woods with the prized meat, which they placed in a sink hole and covered with rocks.

The three hid in the woods with the horses overnight. When they emerged the next morning, they beheld a shocking sight. The figure of an old man was dangling from a tree behind "Bellvue."

David S. Creigh of Greenbrier County had been captured and arrested near his home for allegedly killing a Yankee soldier. Creigh reportedly had discovered a drunken bluecoat molesting his daughter. In the fight that ensued, the Yankee was killed and his body hidden.

While Averell was passing through the area on his way to Staunton, he was told of the incident. Creigh was apprehended and forced to march across the mountains. A "drum head" court martial found him guilty and he was hanged over the protests of several Union officers. Despite warnings not to cut down the body, it was removed as soon as Averell disappeared from view.

Young Colton reported that his grandfather lost 13 horses to the Yankees. In addition, meat left in the smoke house for the next day's breakfast was taken along with the family's entire supply of milk and butter. The crocks containing the diary products were discovered subsequently neatly stacked in a wheat field.[21]

Mollie Patterson and a friend were visiting a family near Jump Mountain when Averell's troopers rode up Walker's Creek. They waited breathlessly, fearing the worst, and were greatly relieved when the Yankees passed on without incident. When Mollie returned to her home, "Mountainview," however, she found that she had not been so fortunate. Gone were all of her flour and bacon. An uncle, David Blackwood, was robbed of all his horses. That evening the Federals feasted on beef and mutton that had been on the hoof that morning.[22]

On the evening of June 10 McCausland visited Francis Smith and Scott Shipp, to whom he revealed information on the Union force approaching Lexington. "John Tiger" explained that he could only delay and harass the larger force. The addition of Captain Thomas E. Jackson's three guns, and the improved arms, ammunition and equipment from the Lexington arsenal would be valuable assets in any forthcoming engagement, he conceded, but not sufficient to spell victory.

During the night the V.M.I. battalion remained under arms. A Cadet howitzer was ordered to the bridge to defend it. Simultaneously, Captain Henry A. Wise led a company of cadets to a nearby hillside in a supporting role, while still other Cadets lined the bridge with turnpentine-soaked bales of hay which, when fired, would assure the destruction of the span.[23]

Hunter's army broke camp early on June 11. Pickets were driven in and McCausland withdrew slowly toward the bridge in East Lexington.

As his men crossed the bridge, "John Tiger" positioned them on bluffs above and below the span. When McCausland's rear guard sprinted for the bridge, emboldened Yankees charged also, trying to cut off the Rebels. A blistering fire from sharpshooters on the bluffs foiled the Yankee strategy and the northerners were forced to fall back on their supports.

Now a Federal battery went into position on an eminence overlooking the bridge and the V.M.I. On the northeast corner of the parade ground, a section of Jackson's battery was primed for action. A brisk artillery duel followed.

As black smoke and flames enveloped the burning bridge, a Federal shell struck the Cadet barracks, showering the youthful defenders with bricks, mortars and splinters. Obviously, this was no place for the Cadet battalion. Officers shifted it into a draw along the road, well away from the barracks. As Federal artilleryists continued their rain of destruction on the Institute buildings, McCausland's weary troopers slowly fell back up the road through Lexington.[24]

The Virginia Military Institute was not the only target for Union artillerists. Private dwellings also were struck. Cornelia McDonald recorded, "Some shells went through the houses, frightening the inhabitants terribly. . . . Our house was struck in several places but no harm done." Like many others, Mrs. McDonald sheltered her children in the basement.

Margaret Preston was more vivid and descriptive in her account of the bombardment. She wrote:

> We have been shelled in reply all day; one shell exploded

ACTION FOUGHT AT
LEXINGTON
June 11, 1864
R. D. Shields
White
Possible routes to
Leyburn's Ford
1st Ky. & 1st Ohio batteries
Crook
Brownsburg
Battery B,
5th U.S. Arty.
Campbell
Leyburn's
displacing to
follow White
after action
Leyburn's Mill
Shaner
McCausland's Retreat
Sullivan
Leyburn's
Ford
Hayes
36th Ohio
(advance
skirmishers)
Hayes' Brigade
crossed at burnt bridge
following action
Bridge Burnt
Magazine
Thomas E. Jackson's guns
Averell
(flanking movement
by way of Rockbridge Baths)
Liberty Hall
(Ruins)
Creek
V.M.I.
"Stono"
To Buena Vista Furn.
& Hamilton's
Cross Roads
"Mulberry
Hill"
Maj. Wise's
Back Creek
McCausland's
Brigade (reformed)
Washington
College
Gov. Letcher's
Spring
Woods
Averell's HQ
after action
Crook's HQ
after action
LEXINGTON
Maj. Paxton's
Col. Massie's
Averell's pursuit
McCoy's
Cadets' retreat
towards Balcony Falls

in our orchard, a few yards beyond us, — our house being just in their range as they threw them at the retreating Confederates. . . . The people from the lower part of town fled from their dwellings, and our house was filled with women and children. Just in the midst of the thickest shelling, the poor wounded boy from the Institute hospital was carried here, surrounded by a guard of Cadets. He had borne the removal well.I have distributed some of J.'s [John Preston's] blackberry wine, which I have always forborne to open, among the frightened and almost fainting ladies.

Captain Russell Hastings, a member of the staff of Union Colonel Rutherford B. Hayes, wrote that "Our artillery. . . sent a few shells through some of the church steeples and the dome of the Military Institute as a gentle reminder."

Achilles J. Tynes, a fatigued officer on McCausland's staff, was sitting on the steps of the Lexington Hotel resting his horse when:

The rascals on the other side of the River. . . got the range of the street completely, & seeing a group of horses & men, let fly a shell among us & so fine was the practice that it passed over our heads about 5 feet, passing up the street about 100 yards beyond us, striking the banister of a porch, rolled harmlessly into the street, striking the porch, of course.

Mrs. William Nelson Pendleton wrote the general:

The wretches shelled the town for hours. Shells fell everywhere in the town — in Colonel Williamson's yard, into Mr. John Campbell's house, into Miss Baxter's house. One struck Colonel Reid's front door and almost struck his daughter A.[gnes]; one fell in the garden here; one struck somewhere near us, as several small bullets were found in the upper porch here and at Captain [David] Moore's.

Still another account flowed from the pen of Fannie Wilson. She remembered:

The first shell that struck our part of the town passed through Mrs. Johnston's house. The next one above ours tearing a circular hole just the size of the ball. We were standing in the front door when the ball passed over our heads, and fearing danger we went to the cellar, thinking that was the safest place, and were standing on the steps when we heard the whizzing near us; we found that another one had passed through our garret wall and struck the rafter, exploding with a thundering noise. It knocked nearly all the plastering off and all the sash out of the windows, made a great many holes in the wall and floor. One piece passed through the ceiling of the passage, two small pieces perforated the ceiling of grandpa's room just above the head of his bed in which he was lying at the time in a doze and was aroused by the fall of the plaster. . . . In a few moments the alarm of fire was given; shells were flying thick and fast. Uncle was the first to run up to the garret and burst the door open where he was almost suffocated by the smoke which was discovered to proceed from the explosion of the shell and not from fire as was supposed. Not long after several white flags were hoisted by some of the citizens when the enemy in turn raised one and the firing ceased.

Dr. William S. White, the Presbyterian pastor, estimated that:

> Twenty houses were struck, some of which were seriously damaged, and two were ignited, but the fire was extinguished. Six shells passed over the parsonage, one exploded in the garden, and one in the stable-yard. But not a person was struck. Many very narrowly escaped.[25]

Warned by McCausland that he could not hold on much longer, Smith ordered Shipp to take the V.M.I. battalion out of town. About 1 p.m. the Cadets departed on the Fairgrounds road. Crossing the North River by a bridge at its mouth, they camped that night near Balcony Falls. The exodus, wrote Cadet John S. Wise, was executed "with heavy hearts. . . through the town, bidding adieu to such of its residents as we had known in happier days. . . . it galled and mortified us that we had been compelled to abandon it without firing a shot."

Although Francis Smith had been able to save much valuable property, the V.M.I. superintendent was forced to sacrifice five guns of the Cadet battery. There simply were no horses available to haul the pieces out of town before the Yankees arrived. As it was, Smith took with him four brass guns and two rifled pieces, complete with caissons and ammunition.[26]

While McCausland's sharpshooters and Jackson's gunners contested the Lexington bridge site, Averell crossed the river at Rockbridge Baths and approached the town on the Kerr's Creek road. Informed of the Yankee's maneuver, "John Tiger" withdrew his command, leaving only a handful of sharpshooters near where the bridge was smouldering. The last grayjacket pulled out of Lexington at about 4 p.m.

Margaret Preston witnessed the arrival of the Northern troops, writing, "The head of the Yankee column came in sight. I went out and watched them approach, saw six of our pickets run ahead of them some ten minutes."

Elizabeth Randolph Preston had another recollection of that moment. Years later, after she had married William Allan, a Confederate ordnance officer, she recalled:

> Mrs. Preston [her stepmother] ran out into the road to warn some men dressed as Confederates that the Federal cavalry was rapidly approaching. One of them turned in his saddle, and with a sardonic grin and a nasal twang, said, "I guess not." They were "Jessie Scouts," Union soldiers dressed as Confederates.

Rose Pendleton, a daughter of the minister-artillerist, sent her father an eye-witness account of the day's happenings. She wrote:

> For two hours there was one continuous stream of cavalry, riding at a fast trot, and several abreast, passing out at the top of town. A quarter after four the vile rabble [Averell's men] came over the hills in swarms and the feeling [of] the poor Lexington people may be better imagined than described.
>
> They advanced into the town on three roads, and came shouting in as though they had captured some valuable post. I must not forget to state after our men passed through three houses hoisted white flags in token of surrender.

To facilitate the Yankees' passage into Lexington, Crook's engineers started to construct a pontoon bridge over the North River. Their progress was too slow to suit some, however. Stragglers and bummers, sensing a golden opportunity to loot, crossed the river on the remnants of the burned bridge and were soon busily at work.

David H. Strother, a Virginian who served as Hunter's chief of staff, rode with the general to the Virginia Military Institute. There he found:

> . . .the sack[ing] already far advanced, soldiers, Negroes, and riffraff disputing over the plunder. . . . The plunderers came out loaded with beds, carpets, cut velvet chairs, mathematical glasses and instruments, stuffed birds, charts, books, papers, arms, cadet uniforms, and hats in most ridiculous confusion. The General stopped at the house of Major [William] Gilham, a professor of the Institute, and told the lady to get out her furniture as he intended to burn the house in the morning. She was eminently ladylike and was troubled, but yet firm. The house was a state building and it was fair to destroy it, yet it was her only home and it was hard to lose it, but she was a soldier's wife and a soldier's daughter so she set us out some good applejack, apologizing she had nothing better, and then went to move out her furniture to the lawn.

Strother and a fellow staff officer visited the home of [Confederate General] Raleigh E. Colston, where they were offered tea.

Resuming his narrative, Strother wrote:

We took our headquarters at the house of Colonel Smith on the lawn of the Institute. In the court of the main building were several pieces of light artillery with a number of limbers and carriages. In front are twelve pieces of bronze cannon of the old French pattern, and there is also a fine copy in bronze of Houdon's statue of Washington.

Gloated a Union cavalryman, "We got all kinds of trophies in the academy."[27]

Captain Russell Hastings remembered: "We were soon over the river into the town marching through the streets singing 'John Brown's body.' We marched through to the other side and went into camp, having made twelve miles. The inhabitants watched us from behind closed blinds, as we were the first Union troops the villagers had seen."

In another reminiscence, a captain in the 18th Connecticut Infantry wrote in his diary:

The Engineer Corps provided a way for us to cross on the ruins of the bridge. The cavalry and artillery cross[ed] at another point. The Engineer Corps used lumber from nearby buildings to make the bridge safe for us to cross. . . . We go into camp on the east side. Lexington is a beautiful town.

Dr. John J. Booth of the 36th Ohio Infantry recorded his impressions of the final resting place of the town's most distinguished hero, Stonewall Jackson:

It is a common grave without head or footstone. A pole stands near the head from which the Rebels hung a flag presented by a gentleman from Liverpool. The boys picked clover leaves from his grave as mementos.

Some Federals cut pieces of wood from Jackson's marker and the flagpole as souvenirs. As one column of Union soldiers passed the grave, they halted, bared their heads and marched slowly and respectfully past.[28]

Mrs. Pendleton and her daughter reported to the clergyman that the northerners "came in by Mulberry Hill, crossing the river near Leyburn's mill — crowds, crowds, crowds."

Margaret Preston — "Miss Maggie" to the townspeople — recorded her personal experiences with the invading horde:

Then the infantry began to pour in: these remained behind, and with cavalry who came in after, flooded the town. They began to pour into our yard and kitchen. I ordered them out of the kitchen, half a dozen at a time, and hesitated not to speak in the most firm and commanding tone to them. At first they were content to receive bacon, two slices apiece; but they soon became insolent; demanded the smokehouse key, and told me they would break the door unless I opened it. I protested against their pillage, and with a score of them surrounding me, with guns in their hands, proceeded to the smokehouse and threw it open, entreating them at the same time, by the respect they had for their wives, mothers, and sisters, to leave me a little meat. They heeded me no more than wild beasts would have done; swore at me; and left me not one piece. Some rushed down the cellar steps, seized the newly churned butter there, and made off. I succeeded in keeping them out of the house. We have had no dinner; managed to procure a little supper; we have nailed up all the windows. I wrote a polite note to Gen. Averell, asking for a guard; none was sent.

Rose Pendleton described conditions at the Episcopalian parsonage:

. . . we were soon overrun by the thieving wretches. We kept the house doors all locked except the front door and while some 2 or 3 of us kept guard there Mama stationed herself with her keys at the back door leading into the kitchen. The first who came asked for something to eat. Ma gave them all the bread and buttermilk she had, and a shoulder of bacon, which had been cooked for Sunday dinner. They were insolent as possible, cursed and swore, vowed they would have anything and everything they wanted, but still Ma was so firm and we showed so plainly that we were not afraid of them, that they were forced to go off not intimidating us. They ran over everything at Col Reid's and Captain Moore's. . . . At Col Reid's they broke into the smokehouse and carried off 1500 lbs of bacon, took from the cellar 5 barrels of flour, all the preserves, butter milk and lard besides bags of servants' clothing which they tore up. At Captain Moore's they went all over the house, though they stole very little, finding scarcely anything. As soon as possible Mama went to the Provost Marshal to demand a guard, but before her return one who had been sent to Captain Moore's rode into our yard, despersed the crowd.

The Pendletons were forced to forgo their evening meal as a result of the servants taking over the kitchen to bake bread for the Yankees.

At the resident of Cornelia McDonald:

We remained as quiet as possible all the afternoon while the town was alive with soldiers plundering and robbing the inhabitants. Some came into our yard, robbed the milkhouse of its contents and passed on their way, picking up everything

> they could use or destroy. . . .
>
> The next morning a squad of men with an officer came to search for provisions and arms. They laughed when on examining the pantry they found only a half barrel of flour and a little tea, all the supplies we had; but their laughter was immense when on ascending to the garret they saw the hens and chickens running over the floor.[29]

When Colonel Strother noticed smoke rising from nearby mountains, General Hunter's aide learned that it was caused by campfires of citizens who had taken refuge in the forest fastness with much of their personal property.

Strother wrote:

> The satisfaction of these people in regard to the Negroes is surprising. They seem to believe firmly that their Negroes will not leave them on any terms. Thus when running off their cattle, horses, and the goods into the mountains, they take their Negroes with them. The Negroes take the first opportunity they find to run into our lines and giving information as to where their masters are hidden and conduct our foragers to their retreats. In this way our supply of cattle had been kept up.

So numerous were the Lexington folks seeking refuge in the mountains that some of their wagons were parked on the roads due to lack of space under cover. The conveyances were easily spotted by Yankee scouts and summarily destroyed. In addition, considerable booty was snatched from the refugees and placed in Union hands.[30]

Depradations continued early on the morning of Sunday, June 12. Maggie Preston, slight of frame but "quick-tempered, impulsive, high-spirited," entered in her diary:

> I slept undisturbed during the night, but was called down stairs early this morning by the servants, who told me the throng of soldiers could not be kept out of the house. I went down and appealed to them as a lone woman who had nobody to protect her. [I] might as well have appealed to bricks. I left the smokehouse door open, to let them see that every piece of meat was taken (I had some hid under the porch, which as yet had not been found). They came into the dining-room, and began to carry away the china, when a young fellow from Philadelphia (he said) took the dishes from them, and made them come out. I told them I was a Northern woman, but confessed I was ashamed of my Northern lineage when I saw them come on such an errand. They demanded to be let in the cellar, and one fellow threatened me with the burning of the house if I

did not give just what they demanded. I said, "Yes, we are at your mercy — burn it down — but I won't give you the key." They then demanded arms; we got the old shotguns and gave them; these they broke up, and left parts of them in the yard; broke into the cellar; carried off a firkin of lard hidden there; a keg of molasses, and whatever they could find; but did not get the bacon. They asked me if we had no more than this: I answered "Yes, but it is in the mountains." Sent to Gen. Crooke for a guard. At last they pressed into the house, and two began to search my dressing room. What they took I don't know. They seized our breakfast, and even snatched the toasted bread and egg that had been begged for the sick man's breakfast. My children were crying for something to eat; I Had nothing to feed them but crackers. . . .

> They carried off the coffee pot and everything they could lay their hands on, and while the guard, a boy of 17, was walking around the house, emptied the corn crib.

Later, the overseer of the Preston farm informed a servant that all of the sheep had been slaughtered and the cattle and horses taken by the Federals.

Conditions at the Pendleton home resembled closely those at the Preston residence. The general's daughter, Susan, recorded:

> . . . a lieutenant and six men came to search. I stepped forward. He came to search for "arms and munitions of war and provisions." I said, "I will show you all I have." "Men, stay," he said. "I should think that one man was enough to go with an unarmed woman," I said. And then we went into every hole and corner. I had part of a barrel of flour and the one wretch broke open Sunday (sic) morning — the guard had made him give back the flour — a water bucketful; . . . also the pieces of bacon, two bushels of meal, and some salt. We went into the empty cellar, into the garret, into all the rooms. He looked, — would not see into the wardrobes — and on the whole behaved well.

The Pendletons recovered their two milk cows, but their ordeal, as well as that of the community, was not yet over. Sallie Alexander Moore remembered a humorous event during the search of the houses: "At Mrs. Compton's, as the officer was going through the house, Miss Lizzie was leading the way upstairs, when suddenly a string broke and a shower of spoons and forks came raining down the steps from under her hoops. The officer was greatly amused, and kindly helped her pick them up and gave them back to her." The day the Yankees left a partridge flew into the house. The bird was captured and eaten that night for supper.[31]

Mrs. Preston reported on June 12 that the despised Yankees "set fire to the Institute about nine o'clock. The flames are now enveloping it; the towers have fallen; the arsenal is exploding as I write."

Among those who witnessed the conflagration was David Strother, who had concurred in Hunter's order to fire the structures. The chief of staff observed the plunderers scampering from the doomed buildings and wrote:

> One fellow had a stuffed gannet from the museum of natural history; others had the high-topped hats of cadet officers, and most of them were loaded with the most useless and impractical articles. Lieutenant [John R.] Meigs came out with fine mathematical instruments, and Dr. Patton followed with a beautiful human skeleton. Some of the officers brought out some beautifully illustrated volumes of natural history which they presented to me. I, however, felt averse to taking anything and left them at Professor Smith's. My only spoil was a new gilt button marked "V.M.I." and a pair of gilt epaulettes which some of the clerks had picked up and handed to me. The burning of the Institute made a grand picture, a vast volume of black smoke rolled above the flames and covered half the horizon. . . . The Institute burnt out about 2 p.m. and the arsenal blew up with a smart explosion. The General seemed to enjoy the scene and turning to me expressed his great satisfaction at having me with him.

Francis S. Reader of the Fifth West Virginia Cavalry added: "We got all kinds of trophies in the academy . . . . There are some of the most extensive libraries here that I ever saw. I have procured some very good works. The cadets who attended the Military Academy here lived in style."

A member of the First New York (Lincoln) Cavalry recorded Hunter's demeanor as he studied the execution of his order. "He stood looking at the burning building, saying as he rubbed his hands and chuckled with delight, 'Doesn't that burn beautifully?' "[32]

With the Institute in ruins, "Black Dave" next turned his arsonic glee to the houses of the faculty members. Strother reported:

> The cottages occupied by Major Gilham and Colonel Williamson were burned. Mrs. Williamson had got her things off. Mrs. Gilham's were all piled on the parade ground and she sat in the midst, firm and ladylike. I asked Prendergast to let her have two wagons and some orderlies to move it. I also got a protection for her at the house she moved into to prevent soldiers from plundering her there.

Strother's self-professed magnanimity was misplaced. Mrs.

McDonald saw Mrs. Williamson and her children scurrying about trying to save their possessions, many of which were lost. Mrs. Gilham spent the night sitting forlornly among her effects in an effort to keep them from the clutches of the avaricious bluecoats. "And not a man dared to help her or offered to take her place," chronicled the refugeeing matron.

While V.M.I. structures succumbed to the flames, an officer handed Hunter a copy of a proclamation by ex-Governor Letcher "inciting the people to arise and wage a guerilla warfare upon the vandal hordes of Yankee invaders," according to Strother. "General Hunter ordered Captain [Matthew] Berry forthwith to burn the Letcher home, allowing the occupants ten minutes to get out of the house. The order was executed without delay."

The Strother account of the sequence of events leading up to the burning of the Letcher home differs slightly from that of Cornelia McDonald, who had been subject to northern barbarities before fleeing Winchester. The 32-year-old diariest reported.

> Mrs. Letcher had consented to entertain two officers at her house, that she had been civilly asked to do. They had spent the night, and eaten breakfast with the family, socially chatting all the while.
>
> When they rose from breakfast, one of them, Capt. Berry, informed Mrs. Letcher that he should immediately set fire to her house. He took a bottle of benzine, or some inflammable fluid, and pouring it on the sofas and curtains in the lower rooms, applied a match and then proceeded up stairs. Mrs. Letcher. . . ran up stairs and, snatching her sleeping baby from the cradle, rushed from the house with it, leaving everything she had to the flames. Lizzie [her oldest daughter] ran up stairs and went into her father's room to secure some of his clothes, and had hung over her arm some of his linen, when Capt. Berry came near her with a lighted match, and set fire to the clothes as they hung on her arm. He then gathered all the family clothing and bedding into a pile in the middle of the room and set fire to them.

Mrs. McDonald was the only person to go to the aid of the stricken lady. "When I reached the scene," she wrote, "Mrs. Letcher was sitting on a stone in the street with the baby on her lap sleeping, and her other children gathered around her. She sat tearless and calm, but it was a pitiable group, sitting there with their burning house for a background to the picture."

Colonel Strother related that "There was quite an excitement around the Letcher house lest the flames should communicate to adjoining houses belonging to his mother, an old woman who used to keep a boarding house. Our soldiers with some difficulty saved it."[33]

Washington College did not escape entirely the fury of the diabolic northerners. When an elderly trustee of the school complained to Strother that the buildings were being ransacked, the officer asked a few questions and then asserted, "I do not wish to discuss the matter, Sir." Then, pointing to the V.M.I. in flames, the hot-tempered Virginian added, "You perceive that we do not intend to discuss it either."

Although the elderly gentleman failed to arouse Strother's sympathies, the college buildings were spared. Hunter, apparently feeling that the college was more of an honor to the "Father of the Country" than a symbol of rebellion, withheld the firebrands.

Which is not to infer that the aged institution escaped damage entirely. The college library, the libraries of the Graham and Washington societies, and that part of the V.M.I. library that had been moved to the adjoining campus for safe-keeping were carried away or destroyed. The chemical and philosophical apparatus of both colleges were destroyed. Not a pane of glass in any of the windows survived, and all the sashes were removed. All closed doors were broken open and all furniture either smashed or stolen. A letter from the college faculty to the Board of Visitors said plaintively, "The whole presents a scene of desolation and destruction which could hardly be surpassed."

Hunter's pyrotechnical outrages were not wholly endorsed by his subordinates. Surgeon Booth wrote that "General Hunter had the Military Institute and Ex. Gov. Letcher's house burned after they had been completely pillaged. He also allowed the Washington College to be gutted.... Its all wrong."

Colonel Rutherford B. Hayes, one of Hunter's subordinates, advised his wife: "Hunter burns the Virginia Military Institute. This does not suit many of us.... Hunter will be as odious as Butler or Pope to the Rebels and not gain our good opinion either." The future president did not hesitate, however, to send a cadet musket to his sons.[34]

Nor was the clergy safe from the vandal's mischief. William S. White, the Presbyterian divine who saw his yard appropriated as headquarters for Averell, snorted:

> They robbed me of my corn and hay worth $500. They cut the curtains from my carriage and carried off a portion of the harness. A well-dressed and well-mounted captain and lieutenant attempted to rob me of my carriage and harness, but were prevented by the sergeant appointed by General Averell to guard my premises.

Despite the theft of his favorite horse, White was charitable toward

Averell, revealing, ". . . [he] acted like a gentleman. He made every practical effort to check the robbers of private property. He stationed a guard at every house where it was asked. . . and punished severely in my presence several men brought before him on a charge of spoiling private houses."

Dr. White's servant John commented on the light-fingered Yankees: "Master, these Yankees are the beat of all the rogues I ever saw, black or white."[35]

Northeners did not vent their wrath on schools and private homes exclusively. Government warehouses at Jordan's Point, containing hay and corn awaiting shipment to Lynchburg, were sacked. The large, valuable Lexington Mills and adjacent buildings met the same fate despite promises they would be spared. A three-story brick factory and other establishments also were fired. The machinery in the woolen mills on Whistle Creek was demolished.

The hospital, too, felt the fury of the despoilers. Medicines, clothing, bedding and food disappeared under the heel of the conqueror.

Proportionately, the less affluent folks of the community suffered more grievously than many of their wealthier neighbors. Francis H. Smith, the superintendent of the V.M.I., disclosed:

> The houses of our poorest operatives, including seamstresses, laundresses, and laborers, were searched, in common with those of the citizens generally, and some of those persons were left in destitute, and almost starving condition. The kindness of friends in Lexington had opened their homes to receive the trunks and effects of cadets. Such houses were made the peculiar objects of vindictive spoilation.
>
> All the regular negro servants of the institution showed a marked fidelity. Our trusty baker, Anderson, the property of the Institute, was stripped of everything, and on being asked whether he had made himself known as belonging to the State, promptly replied, "No, indeed, — if I had told the Yankees that, they would have burnt me up, with the other State property.[36]

Unlike Anderson, many blacks defected to the Yankees as they paraded through Lexington. Noting that many blacks, including her own cook, had participated in the looting of the V.M.I., Mrs. McDonald added, "They all held high carnival. Gen. Crook had his headquarters on a hill near me, in a large handsome house belonging to Mr. [Jacob] Fuller and as it was brilliantly lighted at night and the band playing it was quite a place of resort for the coloured population."

Many Cadet trunks, as well as clothing and equipment, were found subsequently hidden or buried in spots known only to servants. It also was reported that the Fuller home was General Sullivan's headquarters, and that Crook's was at "Mulberry Hill," the estate of Colonel Samuel McD. Reid. Mrs. Preston saved Stonewall Jackson's sword by hiding it in a piano.[37]

Of the numerous atrocities committed by Hunter's hooligans in Rockbridge County, the execution of Matthew X. White, Jr. was likely the most repugnant. White had been a captain in the First Rockbridge Dragoons early in the war. Ill health forced him out of the service in late 1862, but he later enlisted as a private in the Second Rockbridge Dragoons. After hiring a substitute and returning to civilian life, White rejoined the Second Dragoons on June 10. The next morning Confederate pickets, including White, fired on and killed a renegade Virginian who was guiding Hunter's column through Rockbridge County.

Following the capture of Lexington, White returned to his home on personal business. He was reported and arrested. Despite assurances to the contrary by officers staying in the White home, Matt was taken across the river to the Cameron farm and allegedly shot in the back eight times. His body was found by a farmer looking for his cattle. To guard against predators, the Cameron daughters remained all night with the body, which was removed to Lexington for burial the next day.[38]

Looting and pillaging resumed without noticeable abatement on June 13. Owners of the *Gazette* suffered a capital loss when their press and other equipment that had been concealed in the woods were found by Union troops and put out of commission.

Crook's men located and destroyed canal boats loaded with military supplies. Strother remarked, "The boats were burned and a heavy cannonade from the bursting shells which they contained reminded me of a well-contested battle."

One boat was captured a short distance below the mill at the Point. The muskets were smashed over the gunwales of the boat and thrown into the river. Reportedly, a brass cannon rolled overboard from a boat captured at Gooseneck Dam. If so, it was never recovered. Another boat, loaded with the contents of Alexander's warehouse — farm implements, grain, lime and fertilizer — was taken to Loch Laird where it was lowered into a canal lock. The Confederate labors were in vain. The boat was discovered and destroyed. The James River and Kanawha Canal Company saw its carpenter shop, a houseboat and other craft go up in smoke. Hunter reported the capture of six barges loaded with six cannon, ammunition and commissary supplies.[39]

General Duffié's division, which entered Rockbridge County through White's Gap, rode a wave of destruction before arriving in Lexington on the afternoon of June 13. En route, the cavalrymen destroyed the Buena

Vista Furnace and burned the surrounding buildings, the grist and sawmills, the ore washing machine, the trip hammer shop complete with tools, a corn crib with a capacity of 1200 bushels, and 2,000 cords of wood that had been earmarked for charcoal. The horsemen also brought in 80 prisoners and 700 horses.[40]

Hunter delayed his exit from Lexington for a day while awaiting a 200-wagon convoy from which he resupplied his troops with ammunition and rations. Mrs. Gilham, now destitute, was forced to swallow her pride and appeal to Hunter for food. She got it.

The northern general also showed a spark of charity toward the home of Francis Smith. Instead of destroying the house, where he had made his headquarters, as originally intended, "Black Dave" spared the matches when he learned that Mrs. Smith's daughter, who had just given birth, was critically ill.[41]

Averell's division, which departed Lexington early on June 13, encountered McCausland's Brigade near Fancy Hill, in close proximity to the Confederate encampment on Broad Creek. The two forces skirmished hotly before the Rebel cavalry retired slowly with its three guns toward Buchanan. As he had done at the North River bridge in Lexington, "John Tiger" now consigned the bridge over the James River to the flames. Ambushing the Federals with one of his guns, McCausland ordered his "charging squadron" to attack the disorganized Yankees. The tactic succeeded briefly before the two Rebel companies withdrew, and crossed the burning bridge under a hail of bullets. To escape the intense heat on the bridge, McCausland was forced to cross the river on a boat, while some of his men swam to the other side.

Averell's superior numbers steadily pushed back the Rebels, but not without the loss of time to erect a pontoon bridge for the passage of his wagons and artillery. This delay, it was alleged, caused Averell to resort to his own brand of "heat treatment." Eleven buildings in Buchanan were put to the torch. Among those destroyed was "Mount Joy," home of Joseph Reid Anderson, owner of the Tredegar Iron Works in Richmond. Also doomed was the Cloverdale Furnace which, Averell insisted, employed 500 men.

Though McCausland was unable to defeat the Yankees, his delaying tactics were extremely helpful to the Confederate cause. Each obstacle "John Tiger" threw in the progress of the enemy afforded extra time to John C. Breckinridge and Jubal Early, who were en route from Lee's army, to reach Lynchburg.[42].

Early on the morning of June 14 the last Yankee regiments marched out of Lexington, taking with them Hubard's bronze replica of Houdon's statue of Washington, six French cannon and two guns of the Cadet battery.

The three "recording angels" of the community chronicled the exodus of the enemy. Wrote Rose Pendleton:

> . . . the first sound which greeted our ears on Tuesday morning was the tramping of the horses and the rolling of wagons. Before 10 o'clock our own pickets were in town and captured some 2 or 3 Yankees.

Fannie Wilson noted:

> They had been going all night, and by the time we went to breakfast there were not more than a few hundred in the streets. We were in high spirits you may know and there never was so much rejoicing in town.

Margaret Preston penned this account:

> The town began gradually to be cleared, and though we did not know under what rule we were to be considered, we crept out to try to hear something. The experience of our neighbors has been in some instances worse, in some better than ours, but *all* have suffered. Some idea of our absorption of thought may be imagined, when I record that since last Friday till yesterday [Tuesday, June 16] . . . . we actually forgot to eat; four days we went from morning till dark without food.[43]

Few, if any, residents of Lexington suffered greater loss at the hands of the invaders than Miss Maggie's husband, Colonel J. T. L. Preston. She reported: "Mr. P. says that $30,000 would scarcely cover what he has lost . . . . He is a poor man now for the rest of his days, but he bears it with a brave and Christian spirit and offers no complaint."

Preston's fortitude was typical of his Lexington neighbors. The vandals were gone, it was time to adjust, to rebuild their lives and community with the same determination so traditional of their Scotch-Irish forebears.

The *Gazette,* for instance, returned to business shortly after recovering and reconditioning the equipment that had been ruined by the enemy. In an early edition, the editor told of the Union occupation of Rockbridge County:

> During their march through the Valley, they subsisted upon the people. Families were stripped of all their flour, meat, corn, lard, butter and indeed everything laid up for their support. They made no discrimination between the families of Unionists and Secessionists. They plundered all, and declared their purpose [was] to reduce the people to subjugation, by starvation, if their purpose could be accomplished in no other mode. A considerable number of slaves, variously estimated at

> from three to five hundred, were carried off from Rockbridge, about one thousand horses, a large number of cattle, sheep, hogs, and wagons were taken also. The growing crops of grain and grass were considerably damaged, and much fencing destroyed, on their route. The grain cradles, mowing scythes, and farming implements were burned, or otherwise destroyed .... The losses sustained by the people of Rockbridge cannot be less than from two to three millions of dollars.... The people of this County have shown their devotion to the cause in a manner not to be mistaken or misunderstood. In furnishing men and supplies of all kinds, and in the exhibition of their fidelity to the State and the Confederate Government, they have shown as lofty patriotism as those of any portion of the Confederacy .... The enemy themselves declared that the people here — men, women and children, had shown no sympathy for the Union, or for them and that the rebel feeling was stronger here than in any other County they had passed thro'.[44]

While Lexington was free of the captor's tread, other areas of the county were not. Near Natural Bridge, Union scouts spotted deep wagon tracks leading into a pine woods near Greenlee's Ferry. They were tell-tale evidence that Colonel Angus W. McDonald, commandant of Lexington, his son Harry, and Thomas Wilson, together with about a dozen slaves, had sequestered themselves in the forest.

When a part of the First New York (Lincoln) Cavalry approached the camp, it was fired upon. One of its members was knocked from his saddle. The Federals withdrew, only to return later in larger numbers. A demand to surrender was answered with gunfire. The bluecoats attacked the Confederate wagon train, but were driven off until McDonald and his men ran out of ammunition. Then the Yankee cavalrymen divided up the spoils of rifles, shotguns and pistols.

Colonel McDonald received a slight hand wound during the fighting. His wife, Cornelia, recounted details of the engagement:

> At last with oaths and curses they took possession of their prey. One officer, being infuriated at having been kept at bay by two old men and a boy, struck a blow at Mr. Wilson with his sabre, and he fell to the ground with what was to all appearances his death wound. Not satisfied with that, he came near, and stooping, fired his pistol into his temple, so near did he place it, that the ball glancing slightly, took his eye out, and his whole face was filled with the powder.

Miraculously, Wilson survived the ordeal. The two McDonalds were captured and the wagons, after being looted, were destroyed.

The historian of the First New York Cavalry made scant mention of the incident. He recorded that McDonald and several others were captured, along with half a dozen wagons. "The Colonel and his followers," he wrote, "fought bravely, and several of our men were wounded... before they succeeded in taking them. McDonald... was a fine-looking man and a brave soldier."

David Strother, who despised McDonald because of an alleged slight to his father, forced the venerable colonel to walk in handcuffs during the Federal advance to Lynchburg and subsequent retreat into western Virginia.

Harry McDonald soon escaped his captors and fled into the mountains. Wilson, left for dead where he fell, was found the next day by a slave who had been in the fight, and was taken home to recover.[45]

While Hunter was fighting a losing battle at Lynchburg, life in Lexington started to show a faint trace of normalcy. The V.M.I. Cadets returned to town by canal boats on June 25. Finding their barracks in ruins, they took up temporary quarters at Washington College where, at least, they had a roof over their heads. Two days later they were furloughed until September.[46]

Despite Union depredations in the county, the wheat crop was described as fine "with an abundance for home consumption and a large surplus for the army, and for the general market." The corn crop in certain areas was suffering from an extended drought. The Bank of Rockbridge and the Savings Institute lost no time in reopening their doors.

Washington College was converted into a military hospital. When contributions were solicited for the patients, they were furnished swiftly by the ladies of Fairfield and Kerr's Creek, among other communities.

The County Court rallied to the aid of the needy, appropriating another $30,000 for the relief of indigent soldiers' families. William Dold was named agent to distribute the supplies.

In other actions during the summer of 1864, doctors were paid $5,000 for smallpox vaccinations they had given in the past several months and the Lexington town council adopted an ordinance prohibiting blacks from holding secret meetings of the "Benevolent Society." Also banned were religious services unless conducted by a minister or a white member of the church.[47]

Having chased Hunter into the Alleghany Mountains, Breckinridge and Early led their forces down the Shenandoah Valley. Tramping through Lexington, the tatterdemalion veterans passed the grave of Stonewall Jackson where they reversed their arms and bared their heads out of respect for the fallen hero. The townspeople were on hand to greet the ragged and unwashed soldiers and share with them what skimpy foodstuffs they had.[48]

As the crops were harvested in Rockbridge County, a new problem arose — how to get them to market. The problem traced to the mischief of Hunter's minions, who smashed the lock in the canal at Jordan's Point, allowing the water to drain off. Until the bridge over the North River was rebuilt, fording the stream — easy for horses but impossible for wagons — was the only means of going from one section of the county to the other.

While the lock was being repaired, John B. Gibson and David R. Reverly partially rectified the situation by building a ferry boat to haul wagons across the river. Eventually, government engineers arrived to rebuild the bridge, but the project was not completed until near the end of the war. In the meanwhile, the winter's supply of firewood, which ordinarily arrived in Lexington by boat, had to be hauled several extra miles at a significant increase in cost to the inhabitants. The County Court later appropriated funds for the construction of a free ferry.[49]

Left to their own devices because of the widespread shortages, the women of Rockbridge demonstrated a singular ingenuity in keeping their families clothed and fed. Cornelia McDonald was archetypical. Cotton from old mattresses was carded and spun and made into shirts and blouses. Older sons made do with cut-down adult clothing given her by friends.

When a sister in Warren County offered her 50 pounds of wool if Cornelia could send somebody for it, son Harry volunteered to make the 100-mile trip. Three weeks later he returned, carrying the bag on his shoulders most of the distance, when unable to hitch a ride on a passing wagon. Although the woolen mill on Whistle Creek was back in operation, a large portion of its production was for the government. Even that did not hamper the resolute lady. She pled her case so eloquently with the mill superintendent that he consented to card the wool.

Perhaps Mrs. McDonald's dominant triumph was the manufacture of "Confederate candles." These sources of light were made by passing a cord through a pan of melted beeswax. The cord was about six yards long and was repeatedly drawn through till it was as thickly coated as to resemble one's little finger. Bread and potatoes were staples of the kitchen.[50]

As reports filtered in about dwindling Confederate fortunes on other fronts, Rockbridge men over 45 were called out in August, marched to Staunton and organized into two companies. The first of these was led by Captain Charles Kirkpatrick and Lieutenants Samuel P. Wallace and William P. Moore. The other was commanded by Captain Zachariah Kerr and Lieutenants Robert Hutcheson and Robert Montgomery. Captain Freeman's company of 17-year-olds was sworn in for the balance of the war as a cavalry company and assigned to courier duty.

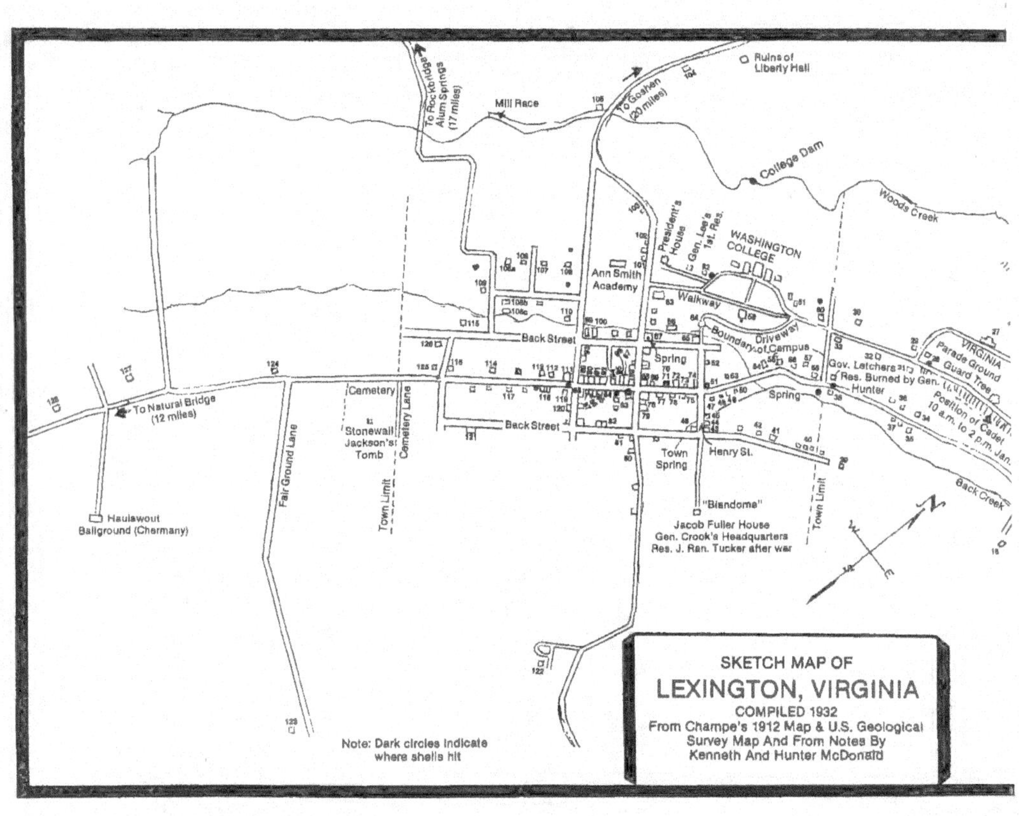

Ruins of Liberty Hall
To Rockbridge Alum Springs (17 miles)
Mill Race
To Goshen (20 miles)
College Dam
Woods Creek
WASHINGTON COLLEGE
President's House
Gen. Lee's 1st. Res.
Ann Smith Academy
Walkway
Driveway
Boundary of Campus
Back Street
Spring
VIRGINIA
Parade Ground
Guard Tree
Gov. Letchers Res. Burned by Gen. Hunter
Spring
Cemetery
Stonewall Jackson's Tomb
Cemetery Lane
Back Street
To Natural Bridge (12 miles)
Fair Ground Lane
Town Limit
Town Spring
Henry St.
Town Limit
Back Creek
"Blandome"
Jacob Fuller House
Gen. Crook's Headquarters
Res. J. Ran. Tucker after war
Note: Dark circles indicate where shells hit
SKETCH MAP OF
LEXINGTON, VIRGINIA
COMPILED 1932
From Champe's 1912 Map & U.S. Geological Survey Map And From Notes By Kenneth And Hunter McDonald

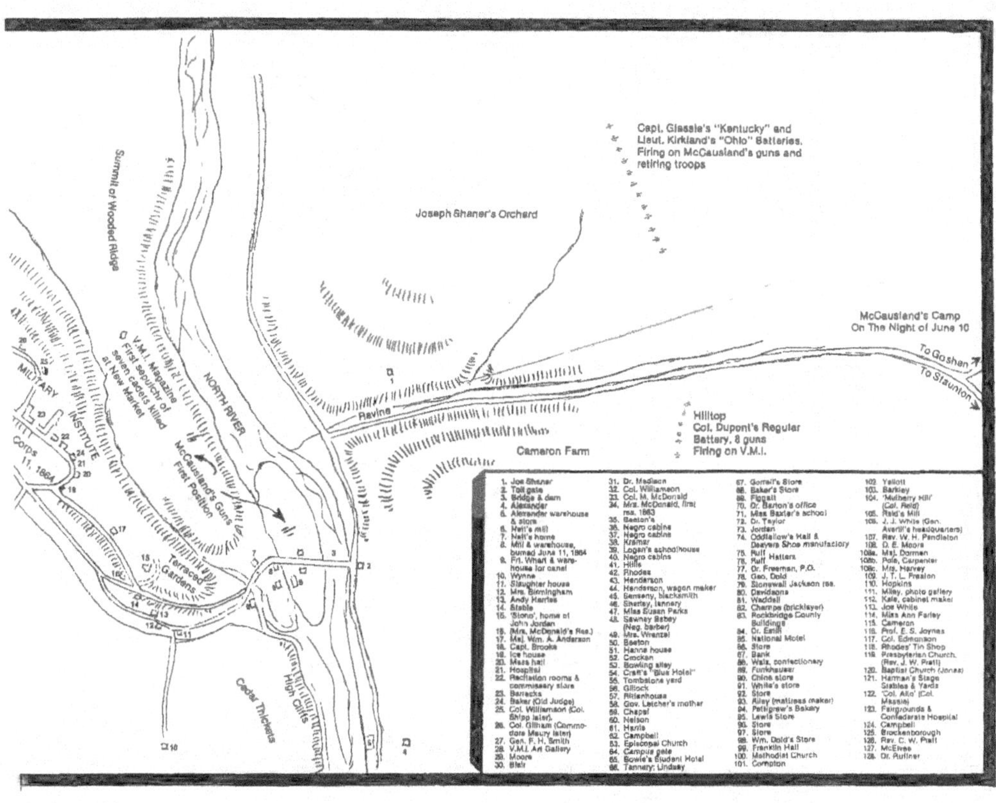
Capt. Glassie's "Kentucky" and
Lieut. Kirkland's "Ohio" Batteries.
Firing on McCausland's guns and
retiring troops
Joseph Shaner's Orchard
Summit of Wooded Ridge
McCausland's Camp
On The Night of June 10
To Goshen
To Staunton
V.M.I. Magazine
First sepulchr of
seven cadets killed
at New Market
NORTH RIVER
Ravine
Hilltop
Col. Dupont's Regular
Battery, 8 guns
Firing on V.M.I.
Cameron Farm
MILITARY INSTITUTE
Corps
11, 1864
McCausland's Guns
First Position
Terraced
Gardens
Cedar Thickets
High Cliffs
1. Joe Shaner
2. Toll gate
3. Bridge & dam
4. Alexander
5. Alexander warehouse & store
6. Neff's mill
7. Neff's home
8. Mill & warehouse, burned June 11, 1864
9. Frt. Wharf & warehouse for canal
10. Wynne
11. Slaughter house
12. Mrs. Birmingham
13. Andy Harries
14. Stable
15. 'Stono', home of John Jordan
16. (Mrs. McDonald's Res.)
17. Maj. Wm. A. Anderson
18. Capt. Brooke
19. Ice house
20. Mess hall
21. Hospital
22. Recitation rooms & commissary store
23. Barracks
24. Baker (Old Judge)
25. Col. Williamson (Col. Shipp later).
26. Col. Gilham (Commodore Maury later)
27. Gen. F. H. Smith
28. V.M.I. Art Gallery
29. Moore
30. Blair
31. Dr. Madison
32. Col. Williamson
33. Col. M. McDonald
34. Mrs. McDonald, first res. 1863
35. Beeton's
36. Negro cabins
37. Negro cabins
38. Kramer
39. Logan's schoolhouse
40. Negro cabins
41. Hills
42. Rhodes
43. Henderson
44. Henderson, wagon maker
45. Senseny, blacksmith
46. Sherley, tannery
47. Miss Susan Parks
48. Sawney Bebey (Neg. barber)
49. Mrs. Wrentzel
50. Beeton
51. Hanna house
52. Crocken
53. Bowling alley
54. Craft's "Blue Hotel"
55. Tombstone yard
56. Gillock
57. Rittenhouse
58. Gov. Letcher's mother
59. Chapel
60. Nelson
61. Harris
62. Campbell
63. Episcopal Church
64. Campus gate
65. Bowie's Student Hotel
66. Tannery; Lindsey
67. Gorrell's Store
68. Baker's Store
69. Flggatt
70. Dr. Barton's office
71. Miss Baxter's school
72. Dr. Taylor
73. Jordan
74. Oddfellow's Hall & Deavers Shoe manufactory
75. Ruff Hatters
76. Ruff
77. Dr. Freeman, P.O.
78. Geo. Dold
79. Stonewall Jackson res.
80. Davidsons
81. Waddell
82. Champe (bricklayer)
83. Rockbridge County Buildings
84. Dr. Estill
85. National Motel
86. Store
87. Bank
88. Walz, confectionary
89. Funkhauser
90. China store
91. White's store
92. Store
93. Alley (mattress maker)
94. Pettigrew's Bakery
95. Lewis Store
96. Store
97. Store
98. Wm. Dold's Store
99. Franklin Hall
100. Methodist Church
101. Compton
102. Yellott
103. Barkley
104. 'Mulberry Hill' (Col. Reid)
105. Reid's Mill
106. J. J. White (Gen. Averill's headquarters)
107. Rev. W. H. Pendleton
108. D. E. Moore
108a. Maj. Dorman
108b. Pole, Carpenter
108c. Mrs. Harvey
109. J. T. L. Preston
110. Hopkins
111. Miley, photo gallery
112. Kale, cabinet maker
113. Joe White
114. Miss Ann Farley
115. Cameron
116. Prof. E. S. Joynes
117. Col. Edmonson
118. Rhodes' Tin Shop
119. Presbyterian Church. (Rev. J. W. Pratt)
120. Baptist Church (Jones)
121. Harman's Stage Stables & Yards
122. 'Col. Alto' (Col. Massie)
123. Fairgrounds & Confederate Hospital
124. Campbell
125. Brockenborough
126. Rev. C. W. Pratt
127. McElvee
128. Dr. Ruffner

In a bid to fill up thinning ranks, Robert E. Lee offered a full pardon to all deserters returning to their commands.

Following Jube Early's decisive repulses at Winchester, Fisher's Hill and Cedar Creek, the conscription laws were tightened even further. Now men between the ages of 17 and 50 were liable for the draft. Farmers and teachers lost their exemption status and most men detailed for specialized work were returned to their commands. A large contingent of these men from Rockbridge County joined the forces of Lee and Early. At approximately the same time, 77 slaves were shipped to Richmond to work on the fortifications.[51]

The ravages of war, crippling as they were, failed to throttle the advancement of education in Rockbridge County the last year of the war. Washington College operated with a sharply reduced student body while the Ann Smith Academy and Mrs. Bull's and Miss Mary Chapin's schools conducted classes in much the same manner as previously.[52]

The citizens of Lexington, still trying to shake off the effects of Hunter's rapacious visit, suffered another shattering blow in late October when word was received confirming the death of a favorite son.

Alexander (Sandie) Swift Pendleton, son of General Pendleton and a graduate of Washington College before his seventeenth birthday, had served on the staffs of Jackson, Richard S. Ewell and most recently that of Jubal Early. At the battle of Fisher's Hill on September 22, Sandie was attempting to restore the center of Early's line when he was shot through the abdomen. He died in Woodstock three days later.

News of Sandie's death was not confirmed until October 7, after communications were reopened between Woodstock and Lexington. The body was returned to the town on October 24 under a military escort commanded by James Power Smith, who had accompanied Jackson's body to its final resting place.

The account of the last rites for the 24-year-old lieutenant colonel was published by the *Gazette*:

> The funeral services were conducted at Grace [Episcopal] Church on yesterday [October 25], in a most appropriate manner by the Rev. Mr. Norton, pastor of the Church, aided by the Rev. Mr. Nelson. The remains of the brave and lamented young officer were conducted to the Lexington cemetery, by the military escort, and a large assemblage of citizens. His resting place is near that of the renowned Jackson, on whose staff he long served with enviable distinction.[53]

Cornelia McDonald also had an extra burden to bear. Learning that husband Angus, the 65-year-old colonel who had been captured some weeks earlier, was about to be exchanged in Richmond, the diarist boarded a canal boat on December 1. At Lynchburg, she transferred to a train and arrived in the capital the next day, only to learn that Angus had died. He was buried in Hollywood Cemetery, with President Jefferson Davis among the mourners.[54]

At various times during the last months of 1864, citizens of Rockbridge were denied their prime news source. Once the *Gazette* missed a publication date through no fault of its own. The paper mill in Mount Solon ceased production and no newsprint was available. At another time, several issues were missed because the river was frozen over and a paper shipment was unable to get through to Lexington.[55]

At such times as the *Gazette* was published, the news generally was disheartening. Sherman's march to the sea and the fall of Savannah, in the eyes of many, augured the death of the Confederacy.

Closer to home, the *Gazette* printed a paragraph about "John Tiger" McCausland:

> In McCausland's late fight in the Valley, his loss was much less than at first rumored. The 14th Va. Cav. sustained the heaviest loss of any part of his command. By the best information that we have, not more than one third of the Regt. was killed or captured, and the loss in it constitutes the most of McCausland's loss. Col. [John A.] Gibson of this county is among the missing and is supposed to have been taken prisoner. This Regt., it is said, fought well, but were overwhelmed.

The wounded colonel and many of his men from Rockbridge County spent the remainder of the war languishing in northern prisons.[56]

Heaping woe upon woe for the Rockbridge residents, a plague of lawlessness broke out in the county during the winter of 1864-1865. Livestock was stolen. Mills, barns and warehouses were levelled by unknown arsonists. Much of the mischief was attributed to deserters and Union men hiding in the mountains. The thieves grew so bold they broke into the office of the Nitre & Mining Bureau in Lexington and made off with several hundred dollars.[57]

In the midst of their gnawing poverty, Rockbridge folks nevertheless responded generously for those less fortunate. Learning that the veterans of the Stonewall Brigade were without shoes, clothing and tents, the Ladies Aid Society of Natural Bridge collected more than $4,000 worth of apparel for the ragged remnants of the once powerful foot cavalry. Other organizations from the county received similar benevolences from the same organization. A committee formed to raise funds for the children of deceased and disabled veterans reported that they had accumulated $20,000 for that purpose.[58]

Christmas of 1864 was a dark and dismal holiday in the county.

Friends of Mrs. McDonald presented her family with a turkey, oysters and cake for dinner. The act of kindness helped Cornelia over a major hurdle, but it was not the answer to her problem. She was down to her last $300 and, after spending that on firewood and other necessities, she was forced to sell her china to hold her family together.

Dinner for the Preston family that winter, wrote daughter Elizabeth, frequently "consisted of a bowl of hot milk poured over toast, or for a change, a bowl of corn meal. . . . Where the paupers were, I do not know. We were all paupers alike."

From the Presbyterian parsonage, Dr. William White gazed out on the wicked ways of the community and observed:

> The whole country seems to be sinking into a state of demoralization. At no time during a ministry of thirty-eight years have I known so much sensual gaiety among professedly pious people, so much drinking of intoxicating liquors, and so free a participation in promiscuous dancing. The present winter . . . has been equally characterized by suffering and sin. Scarcely a family can be found in which death has not recently made inroad. Many families are very scarcely supplied with the commonest comforts. An aged widow told me recently that all the corn she had on which to sustain a family of eight members amounted to ten bushels, for which she paid five hundred dollars. And yet at no period since the settlement of this Valley have there been, in the length of time, as many gay assemblies. Crowds of young people pass from house to house, with little to eat and less to wear, and spend the entire night in dancing and revelry. Sorrow and suffering in themselves uniformly make bad people worse. "The sorrow of the world worketh death." This is divinely true. The state of things is far worse in other sections. . . . A redeeming feature in the case is the readiness with which the people contribute to the support of chaplains in the army. With a few months past more than twenty thousand dollars has been paid into my hands for this purpose by the churches of Lexington Presbytery — an average of eight hundred and seventy dollars to a church.

The congregation of New Providence Presbyterian Church collected $3,000 to purchase artificial limbs for disabled soldiers.[59]

Courtesy: W&LU

John Tilford McKee and James Samuel Mackey. "Liberty Hall Volunteers" Co. I, 4th Inf.

Courtesy: W&LU

B. Gen. Elisha Frank Paxton, Stonewall Brigade. KIA Chancellorsville May 3, 1863.

Courtesy: Mrs. W. W. Heffelinger, Brownsburg, Va.

Surgeon Joseph McClung

Courtesy: Bill Turner, Clinton, Md.

Unknown, 2nd Rockbridge Dragoons, Co. H, 14th Va. Cav.

Courtesy: Bill Turner, Clinton, Md.

Unknown, 2nd Rockbridge Dragoons, Co. H, 14th Va. Cav.

Governor John Letcher

Mrs. John Letcher

Standing left to right: George Washington Shields, Co. I, 4th Va. Inf., James Madison Pettigrew, Co. I, 4th Va. Inf., Rufus Barton McCrum, 1st Rockbridge Artillery, Dudley McDowell (not a veteran). Seated: Joseph White, Lexington Home Guards, Col. John Thomas Lewis Preston, 9th Va. Inf. and V.M.I., John W. Haughawout, Rockbridge Rangers and Senior Reserves and John A. Preston (not a veteran). Circa 1865.

Courtesy: Rockbridge Historical Society

Stonewall Jackson Cemetery

Circa 1866. Note wooden grave markers that no longer exist. Arbor reads "They died for us".

Lexington circa 1866. V.M.I. ruins in background.

Courtesy: Rockbridge Historical Society

Lee-Jackson Camp, CV, Lexington circa 1910. Jeff Shields, Stonewall Jackson's body servant, in center.

Lee-Jackson Camp, CV, Lexington circa 1913

Extreme left front: Chief of Police Parrent (not a veteran), 6th from left David Staples Layne Co. A, 11th Va. Inf. Sitting with paper in his lap Warren Hamilton, unit unknown. 16th from left John Sheridan, Co's C & H, 14th Va. Cav., 14th from left James William Engleman Co. E, 46th Bn. Va. Cav. From right to left from VMI Cadets, 1st J. Ed. Deaver (unit unknown), 3rd John Henry Whitmore, Co. H, 14th Va. Cav., 11th John Adam McNeel, Co. F, 19th Va. Cav.

Courtesy: Rockbridge Historical Society

Lee-Jackson Camp, CV. Lexington circa 1913

1 thru 4 - seated, 5 thru 22 - 1st row, 23 thru 41 - 2nd row. From left: 2. John Edward McCauley, 1st Rockbridge Arty. 3. Samuel Houston Letcher, 1st Rockbridge Arty. 12. Jacob Gassman, Co. F, 7th Va. Cav. 16. James Madison Hayslett, Co. K, 58th Va. Inf. 17. Levi Pultz, Co. C, 1st Va. Cav. 18. Jacob L. Saville, unit unknown. 22. William Clarence Stuart, 1st Rockbridge Arty. 32. John Adam McNeel, Co. F, 19th Va. Cav. (holding flag). 37. James Madison Senseny, Co. H, 27th Va. Inf. 39. John H. Welch, Co. I, 31st Indiana Inf. (U.S.), 40. John Henry Tolley, Co. C, 34th Va. Inf.

Courtesy: Mrs. William W. Thomas, Spottswood, Va.

Confederate Veterans, taken Old Providence Church circa 1913 From left: 1. QM Sgt. George Pilson Lightner, 52nd Va. Inf. 2. Pvt. James W. M. Wallace, Co. E, 52nd Va. Inf. 3. Pvt. J. Martin Harris, Marquis Boys Bty. Va. Res. 4. Pvt. Wm. S. Humphries, Co. E, 5th Va. Inf. 5. Pvt. John P. Smith, 2nd Rockbridge Arty. Standing, 4th from left: William Henry McCormick, Capt. Donald's Co., Provost Guard, Lexington, 5th from left: George W. Wilson, Co. I, 14th Va. Cav., 6th from left: Thomas M. Smiley, Co. D, 5th Va. Inf. Sitting or kneeling: 1. Pvt. David F. Rosen, Co. I, 52nd Va. Inf. 2. Unidentified. 3. Unidentified. 4. From left: Edward A. Demasters, Co. E, 5th Va. Inf., 39th Va. Bn. Cav.

Reunion of survivors, Liberty Hall Volunteers, June 14, 1910 at Unveiling of Lee Table, Washington and Lee University Chapel. Front row from left: Wm. Bell, C. F. Neal, J. N. Lyle, G. B. Strickler, Everard Meade, Wm. Meade, J. L. Sherrard; middle row: H. R. Laird, J. H. B. Jones, Wm. A. Anderson; back row: S. R. Moore, J. P. Amole, Copeland Page, J. T. McKee, T. M. Turner.

Ruins of Glenwood Furnace.

Courtesy: Rockbridge Historical Society

CHAPTER VI

"An End to Valor"

The devaluation of the Confederate dollar grew more acute with every passing day. Dr. White disclosed that "... gold is now worth sixty-five dollars for one in Confederate money, flour twelve hundred dollars a barrel, bacon ten dollars a pound. ... [I] had to pay four thousand seven hundred dollars for a very ordinary horse and cow."

Margaret Preston's family was drinking rye coffee and raspberry tea. In Richmond, Colonel Preston paid $30 for a pair of rubber shoes for his daughter.[1]

On the military front, James J. White was placed in command of the county's armed forces. These consisted of men between 17 and 50 who remained on detail in the area. The Senior and Junior Reserves were currently guarding prisoners and doing courier duty in Richmond. The Fifth Virginia Cavalry and the First and Second Rockbridge Dragoons returned to the county to recruit their mounts. They also assisted the provost marshal and the enrolling officer in rounding up absentees and shirkers. Some members of the First Dragoons participated in General Thomas L. Rosser's raid on Beverly, January 10, 1865.[2]

In political activity, George W. Adams was reelected mayor of Lexington early in the new year. Councilmen were J. M. Pettigrew, Jacob M. Ruff, Robert I. White, William A. Rhodes, Joseph G. Steele and John G. Pole. James M. Paxton was elected sheriff, with John F. Greenlee and David J. Whipple as deputies.[3]

Climatic conditions were extremely harsh the first two months of the year, with temperatures dipping below zero frequently through February. The one positive feature of the bone-chilling conditions was that the ice houses were packed to capacity.[4]

While politicians continued to speak hopefully of peace, those in uniform resolved to continue the war until independence was gained. Members of the Second Rockbridge Artillery urged the homefolks to take heart and continue to support the cause.[5]

Army deserters were appealed to again by General Lee in February. As he had done previously, the commander of all Rebel forces offered a full pardon to all deserters returning to their commands by March 1. His request fell on deaf ears and the Army of Northern Virginia slowly melted away in the trenches around Richmond and Petersburg. The enrolling officer for Rockbridge County also issued a request for all arms to be turned in so they could be sent to the army.[6]

Good and bad news assailed the ears of Rockbridge residents as the severe climate began to moderate. They were thrilled to learn that the bridge over the North River had been rebuilt; they agonized over the news that a spring freshet had broken the canal at Hart's Bottom. As a result of the breakdown in river transportation — and the Federal depredations upon southern communications — the *Gazette* was required to suspend publication for nearly a month.[7]

Food, and the acquisition thereof, remained a dominant thought among the citizenry, particularly after Lee appealed to them for whatever could be spared to feed his famished troops. As usual, Rockbridge responded handsomely and huge quantities of her foodstuffs were shipped eastward.[8]

The arrival of a gentler climate was accompanied by the departure of Rockbridge cavalry units to rejoin the Army of Northern Virginia. With the troopers rode many recruits from the area, most of them 17-year-olds now subject to conscription.[9]

News that Grant had broken Lee's lines south of Petersburg on April 2 delivered a stunning blow to the Rockbridge populace. Even more staggering was the news that arrived a week later — Lee had surrendered to Grant at Appomattox Court House.

Tidings of Lee's capitulation probably were brought into Rockbridge County by Thomas M. Wade, Jr., a member of the First Rockbridge Artillery, who walked home from the surrender site.

Very likely, Lexingtonians got the word from John W. Barclay, who borrowed Wade's copy of Lee's farewell message and rode ahead of Wade into Lexington, where he broadcast the shattering news.

Initially, the first reaction among Rockbridge people was utter disbelief. Then they grew curious and asked questions. Where were the hundreds of sons, brothers and fathers who had remained steadfast to the last? When would they be coming home? The people could only wait, and wait some more — and worry.

The Second Rockbridge Artillery, according to creditable sources, had fought gallantly in defense of Petersburg and had lost most of its officers and men in killed, wounded and captured. The First Rockbridge Artillery, on the other hand, had suffered some casualties on the retreat march, but surrendered the largest company in the Army of Northern Virginia.[10]

The Second Rockbridge Dragoons and the Valley Rangers of the 14th Virginia Cavalry had been at Five Forks and had formed part of the rear guard in the closing scenes of the four-year conflict. At Appomattox on the morning of April 9 they had led a charge that broke the Union line. But that was it. Visible in the distance were long lines of bluecoated infantry, foretelling the doom of Lee's haggard veterans. In the epic charge, the Second Rockbridge Dragoons lost their color bearer, James A. Willson, and Private Samuel A. Walker, both of whom were mortally wounded. Willson died the next day. Walker expired on April 14 and thereby gained the distinction of being the last battlefield casualty from Rockbridge County in Lee's army.[11]

Surrender unleashed another storm of lawlessness in the county. Houses were surrounded by bands of outlaws and deserters, black and white, who broke and took what they wanted. Horses, cattle and sheep were stolen. Armed robbery took place on the roads.

In self defense, citizens organized a police force to patrol the county. Colonel William Gilham was elected chief of the new police force. Jacob Ruff and George Adams were named assistants. Many paroled Confederates joined the law enforcement group and the felons were quickly put to rout. Among those arrested in the crime crackdown were the thieves who had made off with the communion service at the Falling Springs Presbyterian Church.[12]

Without perceptible delay, the ex-Confederates settled into peacetime pursuits. Those with trades returned to their occupations. Many without skills took up farming in order to survive. Despite a shortage of horses and seed, crops were planted. In due time, abundant crops of corn, rye and oats were harvested. It proved a poor year for wheat.[13]

One of Lexington's most prominent citizens found his rehabilitation more difficult than many fellow officers. On his return, William Nelson Pendleton discovered that George H. Norton, who had replaced him as pastor of Grace Episcopal Church, had resigned. "Old Penn" was offered the rectorship on a temporary basis, which allowed him and his family to occupy the manse. Moreover, the impoverished congregation could not afford to pay him a salary. The Pendletons would have to get by on donations from the members.

The former artillerist returned to the pulpit about the time Federal occupation troops arrived in town. In July, a Union officer and an armed guard attended a Sunday morning service. At the close, the officer demanded a copy of Pendleton's sermon. The church was closed for several months, until Pendleton took the oath of allegiance to the United States and Federal troops departed in January 1866. In the spring of that year, the vestry offered Pendleton the pastorate on a permanent basis.[14]

Former Governor John Letcher encountered similar embarrassment in the days immediately following the surrender. On order of General Grant, Letcher and other leading Virginia politicians were placed under arrest. Before daylight on May 20, a cavalry detachment surrounded Letcher's home in Lexington. Major Alexander Moore allowed "Honest

John" a few minutes to pack clothes and bid tearful farewells. Then he was hustled aboard an ambulance that headed down the Valley Turnpike. News of Letcher's arrest flashed ahead of the procession. At various points along the route, the ex-governor was greeted with flowers, food and cheers of encouragement from sympathetic citizens. From Winchester, Letcher was taken to Carroll Prison in Washington. Forty-seven days later, through the intervention of Governor Francis H. Pierpoint of Virginia, Letcher was released. One stipulation was attached to his release. Letcher was required to remain in Rockbridge County and send regular reports of his activities to Washington.[15]

Otherwise, Lexington experienced little interference from occupation forces. Federal authorities cooperated with the mayor and town council in preventing the congregation of idle blacks. The town sergeant was directed to clear the community of all who were unemployed. Those found loitering could be put to work cleaning the streets. To help implement this ordinance, all unlicensed liquor dealers were put out of business.

In other indications that the scars of civil war were fading fast, the stage line between Lexington and Staunton resumed operations, canal boats were arriving at Jordan's Point three times weekly, the railroad was back in service as far as Goshen, and mail was delivered regularly. Mrs. Martha A. Galloway was the new postmistress at Gilmore's Mills.[16]

The first post-election of county officials took place in July. Lucas P. Thompson was elected Judge, John G. Pole, Clerk of the Court; John C. Boude, Assistant Clerk of the Court, and James K. Edmondson, John H. Moore and Charles H. Davidson, Commissioners in Chancery. Edmondson, who lost an arm at Chancellorsville, wrote his wife about the voting:

> I came into town . . . with a majority from the two precincts of [Natural] Bridge and Dryden's School house of about 250. The Bushwhackers and the union men however of town and House Mountain and the tail end of Collierstown had formed a combination and came near carrying the town polls. . . . my friends however rallied to the rescue and carried the polls about 30 majority. . . . A general row took place at the Bridge in the afternoon of the day of election. It was instigated by the Bushwhackers and deserters. They were finally driven off, no serious damage done, the rocks flew thick and fast and by dint of perseverance in the dodging line, I saved my noggin from being bruised.

David S. G. Cabell represented Rockbridge County in the Virginia Senate, while Doctors Alexander Graham and James McD. Taylor did so in the House. William D. McCorkle was elected sheriff.[17]

Also in July, John Brockenbrough reopened his law school. Later, Washington College and the Virginia Military Institute resumed classes. Other educational institutions reopening in the county included the Ann Smith Academy for girls, the Brownsburg Classical School for boys, Miss Bull's School, and a new English and Classical School at Fancy Hill under the direction of William T. Poague and David E. Laird.[18]

On September 18, 1865 an elderly gentleman rode into Lexington on a steel gray horse. When he dismounted at an inn, he was immediately recognized and surrounded by old soldiers. Robert E. Lee had arrived in town! Some weeks before, Judge John Brockenbrough, rector of Washington College, had visited the general at his temporary home in Cumberland County and notified him of his election to the presidency of the college by the Board of Trustees. Lee, who had rejected many more lucrative offers of employment, accepted Brockenbrough's offer because Washington College was a private institution and non-sectarian.

For four days, Lee rode "Traveller," his long-time companion, across mountains and valleys and arrived in Lexington a day before he was expected. Fortunately, James J. White, a professor at the college, happened by. He escorted Lee to the home of Colonel Samuel McD. Reid, where he was welcomed warmly. Noted the *Gazette:* "Our modest unpretending little town would glady have made some general demonstration of welcome to the venerable hero, if we believed that he would have relished anything of that kind, and if he had given us an opportunity." Lexington was to be Lee's home until the end of his days.[19]

The War for Southern Independence may have become the "Lost Cause," but the people of Rockbridge did not forget the widows and orphans of men who had fought in their behalf. Contributions were collected throughout the county for their support.

The Lexington town council forgave John Letcher's unpaid taxes on his home that had been burned, as well as the John Gibson & Company property at Jordan's Point that also had been destroyed. James S. Smith filed a petition for exemption from his delinquent taxes of 1864, explaining that he was in the army that year, and furthermore, he was unable to pay his debt with Confederate currency. The council acted favorably on his request.

Patrols, consisting of one captain and ten men, continued to provide night-time security for the citizens. The cost of the candles used in the program was paid for by the town.[20]

Signs of economic growth in Rockbridge County were abundant that year of 1866. One example was the two pages of advertisements, offering goods and services, that appeared in the *Gazette*. Another indication of recovery was Andrew W. Varner, who was busily engaged in making furniture for General Lee's house on the Washington College campus. A popular topic of discussion on the streets was the ten thousand dollar gift to the college by Cyrus McCormick, a Rockbridge native who had made a fortune as inventor of the reaper. A Mr. Wilson was another benefactor. The Philadelphian donated his valuable library to the school. Citizens also were urged to return to the town's two colleges any books that might be found around the county. Two popular pre-war resorts, Rockbridge Alum Springs and the Rockbridge Baths reopened, another unmistakable sign that better times had come.[21]

The rebirth of the economy did not blot out memories of fallen heroes. Henry Kyd Douglas, who had been on Jackson's staff, launched a campaign to raise funds for an equestrian statue over Old Jack's grave. A. Frederick Volck was to design and execute the memorial. The bodies of the ten Cadets who had been killed at New Market were returned to Lexington and reinterred at the Virginia Military Institute. The ladies of Lexington, taking a cue from their counterparts in Winchester, arranged for the bodies of the soldiers buried at the Fairgrounds to be removed to the Presbyterian, now Stonewall Jackson Memorial, Cemetery.

On the third anniversary of Jackson's death, an impressive service was conducted at the Presbyterian Church led by Reverend General Pendleton and Reverend Doctor White. From the church a procession filed to the cemetery, led by veterans of the Stonewall Brigade, followed by the Cadets, citizens and ladies of the town, and students at Washington College. Instead of a wooden marker over the grave, there was now a headstone engraved simply: "Lieut. Gen. Jackson." As reported by the *Gazette:*

> He has no epitaph and needs none. Angel hands have recorded it in Heaven — his noble services — his purity of life, and his heroic Christian death.
>
> It certainly was an impressive and beautiful sight, to look upon the large congregation of persons there assembled, of every rank and condition of life, and of every age, who, in thus honoring the dead, honored themselves. Among them was the leader of the Southern armies, who stood with uncovered head, as he tendered silent homage to the heroic sleeper.
>
> The heroic Chief is surrounded by his heroic followers in death as in life.
>
> The knights are dust,
> Their good swords are rust,
> Their souls are with the saints, we trust.[22]

EPILOGUE

In the years immediately following the Civil War, interest in the events and personalities of the conflict was limited to occasional mention in the newspapers. Stonewall Jackson's grave and those of the Confederate soldiers buried in the Presbyterian Cemetery continued to be decorated by the ladies of the community. The Virginia Military Institute Cadets and the students of Washington College (later Washington and Lee University) participated in these stirring ceremonies. The veterans turned out in large numbers for these events. Eulogies for the fallen heroes, such as Jackson, and later Lee, appeared from time to time in the local press. The longest eulogy printed in the Lexington paper was that of a relatively unknown hero, Lieutenant Colonel Thomas H. Watkins of the 52nd Virginia Infantry, killed while leading his regiment in the battle of Bethesda Church in 1864. Many soldiers passed away without an obituary and/or mention of their military service. In the 1870's reunions of some units, such as the Stonewall Brigade, were held at the Augusta County Fair. In 1875 Jackson's sturdy veterans and the First Rockbridge Dragoons journeyed to Richmond for a reunion at the Virginia State Fair. Company rosters of units from Rockbridge County and other nearby areas appeared in the newspapers from time to time.

In 1884 the survivors of the First Rockbridge Artillery conducted weekly meetings at the firehouse in Lexington. A committee was formed in the county to honor the deceased boys in gray on Confederate Memorial Day. Subscriptions and donations poured in for the Soldiers Picnic to be held after the ceremonies in the cemetery. This proved to be a huge success and was continued over the years, although later altered to a dinner, as the years diminished their numbers. At the unveiling of Lee's recumbent statue on June 26, 1884, an estimated 5,000 people attended. The old gunners of the First Rockbridge Artillery fired several salutes from the original guns of the battery. This event sparked the interest of the old soldiers from other units, and Lieutenant Colonel Watkins' original company, E, of the 52nd, from the Natural Bridge area, met at the county courthouse during June, followed by the First Rockbridge Dragoons in August. Members of the Fifth Virginia Infantry trekked to Niagara Falls, N.Y., to meet with members of a Union regiment they had faced in battle. On July 21 the first reunion of the First Rockbridge Artillery was held in Lexington.

In 1888 the Stonewall Jackson Camp of Confederate Veterans was organized in Lexington, uniting all the old soldiers in the county into one group. Monthly meetings were held at the courthouse. The camp passed judgment on requests for pensions by former Confederates. In a few cases these applications were disapproved, always because of desertion. In later years the pensions were generally approved, past crimes being forgotten or unknown to the officers of the camp. In July, 1888, members of the Fourth Virginia Infantry traveled to Gettysburg for a reunion with Union veterans they had faced across the battle lines. In 1890 many of the members of the camp attended the Confederate reunion in Richmond. The First Rockbridge Dragoons also met in Lexington.

During 1891 the First Rockbridge Artillery, First Dragoons and the Rockbridge Rifles of the 27th Virginia Infantry held reunions in Lexington. Following the meetings the veterans marched to the cemetery for the unveiling of the statue of Stonewall Jackson. The old cannoneers again fired their guns in honor of their departed leader. The survivors of the Stonewall Battery and the Stonewall Brigade met in Staunton a few days later. Despite this outpouring of enthusiasm, the veterans camp apparently dissolved. It was revived in 1895. Meanwhile, the gray riders of the 14th Virginia Cavalry held a reunion in Lexington in 1894.

Perhaps spurred by the revival of the Stonewall Jackson Camp, the United Daughters of the Confederacy formed a Chapter in Lexington during 1896. They were the force behind the Lee-Jackson Day dinners for the old soldiers in future years. While company, regiment and brigade reunions continued for a while, the veterans camps became the focal point for the old Johnnies. In 1901 the Sons of Confederate Veterans chapter was formed in Lexington. This group grew to over 200 members before disbanding about 1949.

The First Rockbridge Artillery fired its guns for the last time in 1913 at the 50th anniversary of Stonewall Jackson's death. The old battery members met for the last time as a group in Richmond, at the 1915 Confederate reunion. World War I saw the revival of the Rockbridge Artillery and it served with distinguished valor in France. The unit continued to serve in the Virginia National Guard after the war.

As time reduced the ranks of the old soldiers the attendance at the dinners sponsored by the U.D.C. decreased. In 1935 only three veterans remained in the county; J. A. Mankey, John D. McNamara and Levi Pultz. Pultz survived until March 2, 1936, when he passed away at his home on Timber Ridge. Appropriately, he was buried in Stonewall Jackson Memorial Cemetery, Rockbridge's last Confederate soldier. Benjamin M. Cash, a native of Augusta County, who served in a Nelson County unit, died in Brownsburg on February 5, 1945, age 98 years, the last veteran in the county.

APPENDIX A

CONFEDERATE CEMETERY, ROCKBRIDGE ALUM SPRINGS

There are approximately 52 Confederate soldiers buried at the Rockbridge Alum Springs. The resort was used as a Confederate hospital during the war. There are no markers except field stones. Listed below are some of the Confederate soldiers known to be buried there.

Private WILLIAM G. THOMPSON, Company E, 14th Georgia Infantry. Died there October 16, 1861.

Private WILLIAM COX, Company E, 14th Georgia Infantry. Died there October 20, 1861.

Private OLIVER P. ORR, Company E, 14th Georgia Infantry. Died there October 21, 1861.

Private JOHN E. PERKINS, Company A, 14th Georgia Infantry. Died there November 28, 1861.

Private DAVID L. DAVIDSON, Company C, 14th Georgia Infantry. Died there 1861, exact date unknown.

APPENDIX B
STONEWALL JACKSON MEMORIAL CEMETERY

Burials from the Fairview Confederate Hospital at the Fairgrounds, Lexington, and removed to Stonewall Jackson Memorial Cemetery in 1866, and other burials in the Confederate Section and two unknown Confederate soldiers buried elsewhere in the cemetery.

A. L. Killian	Co. A 23rd North Carolina Inf.	d. 2/3/63
Willis G. Gilbert	Co. D 25th North Carolina Inf.	d. 2/6/63
D. Speight	Co. F 41st Virginia Inf.	d. 2/26/63
J. T. Perry	Co. B 10th Alabama Inf.	d. 3/7/63
H. H. Cooper	Co. C 3rd South Carolina Bn. Inf.	d. 3/8/63
B. L. Byrd	Co. A 6th Virginia Inf.	d. 3/10/63
Charles Todd	Co. B 26th Georgia Inf.	d. 3/26/63
W. H. Walters	Co. G 2nd North Carolina Inf.	d. 4/7/63
B. Matheny	Co. G 2nd Virginia Infantry	d. 4/19/63
J. C. McKinney	Co. B 34th North Carolina Inf.	d. 6/16/63
H. D. Bryant	Co. A 7th South Carolina Inf.	d. 7/13/63
W. H. Johnston	Co. D 28th Virginia Inf.	d. unknown
Captain J. C. Day	Co. H 5th Tennessee Cavalry	KIA 10/6/64
H. B. Huggins	Co. I 2nd North Carolina Inf.	d. 6/16/63
J. B. Byrd	Co. E 31st Virginia Inf.	d. 12/26/63
W. W. Everly	Co. __ 23rd Virginia Inf.	d. unknown
James M. Campbell	Co. H 27th Virginia Inf.	d. 2/23/62
Nathan O. Hook	Co. H 27th Virginia Inf.	After 1870
Robert Emmett McAleer	Co. H 27th Virginia Inf.	d. 3/18//79
David Guthrie Bowyer	Co. H 27th Virginia Inf.	d. 3/14/98
Joseph A. Higgins	Co. H 27th Virginia Inf.	d. 10/5/98
Lucas P. Thompson	1st Rockbridge Artillery	1835-1925

There are many more Confederate soldiers buried in this cemetery, most of whom are not identified as such.

INTRODUCTION TO THE ROSTERS

The muster rolls are not complete. The rosters contained in the unit histories in the Virginia Regimental Series, whose rolls were made up from the Compiled Service Records in the National Archives, are the primary source for the lists. The quality of these service records varies from unit to unit and all records end several months or more before the war was over. Companies mustered for pay every two months. However, some of these records do not exist, especially toward the end of the war. The Roster of Companies of Rockbridge County Men in the War Between the States, in the Rockbridge County Courthouse (copied in Morton's History of Rockbridge County) was a valuable addition. However, this book was made up circa 1900, relying on the memories of the old soldiers, therefore omitting numerous men from the organizations from the county. The County Death Records, Marriage Bonds, Order and Minute Books, and other Civil War papers in the courthouse, added additional names. The *List of Exempts, Reserves and Substitutes from Rockbridge County,* in the Rockbridge Historical Society Collection, provided a vast amount of data, including descriptive lists of the men. The militia muster rolls, veterans papers and other rosters and lists given to the Society over the years, were other valuable sources. The Virginia State Library contains rosters of most Rockbridge units, muster rolls and postwar rosters of some units, pension applications, and other data that provided additional names for the rolls. The 1860 census gave the ages, occupations, and locations of most of the males in the county. The 1870 census did the same, and confirmed that the individual had survived the war, in most cases. The 1910 census identified most of the Union and Confederate veterans alive in the county. Family genealogies, cemetery records, United Daughters of the Confederacy applications, and many other sources were used to identify Rockbridge soldiers.

There is duplication of names. Many soldiers served in one or more units and a few served in four or five different ones. The Liberty Hall Volunteers and the 1st Rockbridge Artillery both contained numerous men from other counties. In order to give complete muster rolls of these units, and other ones, all the names are included. All soldiers who were living in Rockbridge County in the 1860 census, resided here after the war, or died in the county are included. Some men who moved to other areas before the war were also added. A few men from the county enlisted in the Federal service and are included. Confederates who took the oath of allegiance and enlisted in the Union army are identified only in the Southern unit. The newspaper printed rosters of the companies from time to time, and obituaries were major sources of identification for many men who were not otherwise known to have served. Additions to these rosters will be gratefully accepted.

ABBREVIATIONS

Term	Abbreviation
Adjutant	Adj.
Aide-de-camp	ADC
Artillery	Arty.
Assistant Adjutant General	AAJ
Assistant Surgeon	Asst. Surgeon
Battalion	Bn.
Battery	Bty.
Brigade	Brig.
Brigadier General	Brig. Gen.
Captain	Capt.
Cavalry	Cav.
Colonel	Col.
Commissary Officer	Comm. Off.
Company	Co.
Confederate States Navy	CSN
Confederate States Ship	CSS
Corporal	Cpl.
Division	Div.
General	Gen.
Infantry	Inf.
Lieutenant	Lt.
Lieutenant Colonel	Lt. Col.
Major	Maj.
Ordnance Officer	Ord. Off.
Quarter Master	QM
Regiment	Regt.
Sergeant	Sgt.
Sergeant Major	Sgt. Maj.
Surgeon	Surg.

FIRST ROCKBRIDGE ARTILLERY

The First Rockbridge Artillery was organized in Lexington on April 21, 1861. The original company was composed primarily of men from Lexington and Rockbridge County. The battery continued to receive recruits from Rockbridge County and throughout the state, many of whom were college graduates or students. For complete information on this unit and the men who served in it, see Robert J. Driver's, *1st and 2nd Rockbridge Artillery*, 1987.

Brigadier Generals
John McCausland and William Nelson Pendleton
Lieutenant Colonels
William McLaughlin and William Thomas Poague

ADAMS, THOMAS T.
AGNER, AUGUSTUS W.
AGNER, JOHNATHAN
AGNER, JOSEPH SPRIGGS
AGNER, OSCAR WILLIAM
ALEXANDER, EDGAR S.
ALEXANDER, JOHN McDOWELL: 2nd Lt.
ANDERSON, JOHN
ARMISTEAD, CHARLES JAMES
ARNOLD, ABNER E.: Surgeon
AUSTIN, JOHN
AYRES, JAMES J.
AYRES, SAMUEL ROGERS
BACON, EDLOE PHILIP, JR.
BANE, ANDREW
BANNISTER, ISAAC M.
BARNETT, JOHN EDWARD
BARTON, ROBERT THOMAS
BEARD, JOHN R.
BEDINGER, GEORGE RUST
BERRY, JAMES B.
BLACK, JAMES
BLACKFORD, LANCELOT MINOR
BOLLING, WILLIAM HOLT
BOTELER, CHARLES PEALE
BOWYER, JOHN H. A.
BROCKENBROUGH, JOHN BOWYER: 1st Lt.
BROOKE, PENDLETON
BROWN, HENRY CLAY
BROWN, JOHN MATHEWS
BROWN, WILLIAM MORTON
BULL, ROBERT S.
BURWELL, PHILIP LEWIS
BYRD, WILLIAM
CALLOMY, J.
CARUTHERS, THORNTON R.
CHILDRESS, WILLIAM A.
CLARK, JAMES GIBSON
COCHRAN, FIELDING J.
COFFMAN, JOHN H.
CONNER, ALEXANDER
CONNER, FITZALLEN Y.
CONNER, HENRY CLAY
CONNER, JOHN C.
COOKE, RICHARD D.
CRAIG, JOHN B.
CROSON, WILLIAM G.
DANDRIDGE, ADAM STEPHEN
DARNELL, HENRY THOMAS
DAVIS, JAMES COLE: 2nd Lt.
DAVIS, JOHN E.
DAVIS, RICHARD G.
DIXON, WILLIAM HENRY HARRISON

ADKINS, BLACKBURN
AGNER, JOHN D.
AGNER, JOHNATHAN T.
AGNER, McD.
AGNER, SAMUEL S.
ALEXANDER, EUGENE
ALVIS, THOMAS T.
ANDERSON, SAMUEL D.
ARMSTRONG, C. J.
ARNOLD, ALBERT
AYERS, G. N.
AYRES, NAPOLEON BONAPARTE
BACON, EDLOE PHILIP, SR.
BALDWIN, WILLIAM LUDWELL
BANE, SAMUEL R.
BARGER, WILLIAM G.
BARTON, DAVID RITTENHOUSE
BEALL, JESSE T.
BEARD, WILLIAM B.
BELL, ROBERT SHERRARD, JR.
BLACK, BENJAMIN FRANKLIN
BLACK, JAMES C. FRANKLIN
BLAIN, DANIEL
BOTELER, ALEXANDER ROBINSON, JR.
BOTELER, HENRY
BOYD, ELISHA HOLMES
BROCKENBROUGH, WILLOUGHBY NEWTON
BROWN, E.
BROWN, JOHN L.
BROWN, RICHARD W.
BRYAN, EDMUND
BUMPUS, WILLIAM N., JR.
BYERS, GEORGE NEWTON
BYRD, WILLIAM M.
CARSON, WILLIAM
CHAPIN, WILLIAM TAYLOR
CHRISTIAN, M.
CLARK, JAMES GREGORY
COFFEE, ABNER WHITFIELD
COMPTON, ROBERT KING
CONNER, DANIEL
CONNER, GEORGE W.
CONNER, JAMES A.
CONNER, ROBERT BLANE
COX, WILLIAM H.
CROCKEN, FRANCIS JAMES P.
CURRAN, DANIEL
DARNELL, ANDREW M.
DAVIS, CHARLES WARREN
DAVIS, JAMES M. M.
DAVIS, MARK
DICKEY, R. D.
DODD. ROBERT DUNBAR

DOLD, CALVIN MORGAN
DOUGLASS, WILLIAM WALTON: Surgeon
DUDLEY, ROBERT M.
EFFINGER, WILLIAM HENRY
EPPES, W. H.
FAIRFAX, RANDOLPH CARY
FIGGATT, FERDINAND W.
FONTS, HENRY
FORD, JAMES A.
FRIEND, BENJAMIN CARTER MINGE
GARNETT, JAMES MERCER
GAY, HENRY ERSKINE
GIBBS, JOHN TRACY, JR.
GIBSON, JOHN THOMAS
GILLIAM, WILLIAM T.
GILMER, JAMES BLAIR
GINGER, GEORGE ALEXANDER
GOLD, ALFRED
GOOCH, JAMES T.
GORDON, WILLIAM C.
GRAHAM, ARCHIBALD A., JR.: Captain
GRAY, THOMAS P.
GREEN, THOMAS
GREGORY, JOHNATHAN MUNFORD, JR.
HALL, JOHN F.
HAMILTON, AUGUSTUS HOUSTON
HARRIS, BOLLING
HEISKELL, JACOB CAMPBELL
HERNDON, BRODIE STRACHAN, JR.: Surgeon
HETTERICK, FERDINAND
HITNER, JOHN KENNEDY
HOLMES, JOHN ANTHONY
HOSTETTER, GEORGE WILLIAM
HOUSTON, WILLIAM WILSON
HOWERTON, JOHN C.
HUMMERICKHOUSE, JOHN R.
HYDE, EDWARD H.
JOHNSON, LAWSON W.
JOHNSTON, WILLIAM FINLEY
JORDAN, JOHN WILLIAM, JR.: 2nd Lt.
KEAN, OTHO GARLAND
KNICK, WILLIAM
LACY, RICHARD B.
LANE, JOHN S.
LAWSON, WILLIAM
LEE, JAMES W.
LEECH, JAMES M.
LETCHER, SAMUEL HOUSTON
LEWIS, JAMES PAYNE
LEWIS, ROBERT PAYNE
LINK, DAVID
LOWMAN, JAMES WILLIAM
LUKE, WILLIAMSON
McALPIN, ROBERT
McCAULEY, JOHN EDWARD
McCLINTIC, WILLIAM SHIELDS
McCORKLE, JOHN BAXTER: 2nd Lt.
McCORKLE, THOMAS EDWARD
McCORKLE, WILLIAM ALEXANDER
McGUIRE, HUGH HOLMES, JR.
MACON, LYTTLETON SAVAGE
MAGRUDER, HORATIO ERSKINE
MARSHALL, OSCAR M.
MASSIE, JOHN LIVINGSTON
MAUPIN, CHAPMAN
MAURY, THOMPSON BROOKE
MEADE, FRANCIS ALEXANDER
MICHAEL, BENJAMIN F.
MINOR, CHARLES BERKELEY
MINOR, LANCELOT
MONTGOMERY, WILLIAM GRAHAM
MOORE, DAVID EVANS, JR.
MOORE, JOHNATHAN D.
MOORE, JOHN HARVEY
MOORE, JOHN POAGE
MOORE, WILLIAM DORMAN
MORGAN, GEORGE W.
DOREN, JOHN
DOWAY, W. D.
DUNLAP, SAMUEL McKEE
EMMETT, MICHAEL J.
ESTILL, WILLIAM C.
FAULKNER, ELISHA BOYD
FISHBURNE, CLEMENT DANIELS
FORD, HENRY F.
FRAZER, ROBERT
FULLER, JOHNATHAN
GAY, CHARLES WYNHAM
GERALD, EDWARD
GIBSON, HENRY B.
GIBSON, ROBERT ATKINSON R.
GILMER, GEORGE HUDSON
GILMORE, JAMES HARVEY
GINGER, WILLIAM LEWIS
GOLD, JOHN M.
GOODWIN, W. A.
GOULD, JOHN M.
GRAHAM, JOHN ALEXANDER: Surgeon
GREEN, JOHNATHAN
GREEN, ZACHARIAH
GROSCH, CHARLES
HAMILTON, ANDREW JACKSON
HARRIS, ALEXANDER
HEISKELL, ISAAC PANCAKE
HENRY, NORBORNE SAMUEL
HERNDON, FRANCIS T.
HICKS, WILLIAM
HOGPADE, W.
HOLTZCLAW, E. TAYLOR
HOUSTON, JAMES RUTHERFORD
HOUCHENS, J. C.
HUGHES, WILLIAM
HUTTON, GARDNER PAXTON
JAMES, SAMUEL
JOHNSON, THOMAS E.
JONES, BEVERLY RANDOLPH
JORDAN, WILLIAM C.
KEAN, WILLIAM CALLIS
LACKEY, JOHN H.
LACY, WILLIAM STERLING
LAWSON, JOSEPH
LEATHERS, JOHN PENDLETON
LEE, ROBERT EDWARD, Jr.
LEOPARD, JAMES N.
LEWIS, HENRY PAYNE
LEWIS, NICHOLAS HUNTER
LEYBURN, JOHN: 2nd Lt.
LOTTS, JAMES FRANKLIN
LLOYD, L. MARSHALL
McALPHIN, JOSEPH M.
McCAMPBELL, DAVID AUGUSTUS
McCAULEY, WILLIAM H.
McCLURER, JOHN GRIGSBY
McCORKLE, TAZEWELL MORTON
McCORKLE, WILLIAM A.
McCRUM, RUFUS BARTON
McKIM, ROBERT BRECKINRIDGE
MAGRUDER, DAVENPORT NEAL
MARSHALL, JOHN J.
MARTIN, THOMAS
MATEER, SAMUEL LINDSAY
MAURY, MAGRUDER
MAY, BUSHROD LAFAYETTE: Surgeon
MERRICK, ALFRED D.
MINOR, CARTER NELSON BERKELEY
MINOR, CHARLES LANDON CARTER II
MONTGOMERY, BENJAMIN D.
MOORE, A. B.
MOORE, EDWARD ALEXANDER
MOORE, JOHN F.
MOORE, JOHN LYLE
MOORE, SAMUEL R., JR.
MOORE, WILLIAM MARTIN VAN BUREN
MUTERSPAUGH, WILLIAM M.

MYERS, JOHN McN.
NELSON, KINLOCH
NICELY, GEORGE H.
NICELY, JOHNATHAN F.
OTEY, WILLIAM NEWTON MERCER
PACKARD, JOSEPH, JR.
PAGE, WILLIAM CHANNING MOORE
PAINE, DAVID BRAINARD
PAINE, HENRY RUFFNER
PAINE, WILLIAM H.
PATTERSON, JOHN C.
PAXTON, JAMES LEWIS
PAXTON, SAMUEL L.
PENDLETON, ALEXANDER SWIFT
PENDLETON, EDMUND
PHILLIPS, JAMES H.
POLLARD, JAMES G., Jr.
PORTER, ______
PRESTON, FRANKLIN C.
PUGH, JOHN A.
RADER, DANIEL P.
RAWLINGS, JAMES MINOR
REINTZEL, GEORGE W.
RHODES, JACOB N.
ROBINSON, ARTHUR
ROOT, ERASTUS COLIN
RUTLEDGE, CHARLES A.
SAVILLE, JOHN
SCHMIDT, ADAM
SHARP, BENJAMIN F.
SHIRLEY, RICHARD
SILVEY, JAMES HENRY
SMITH, JAMES MORRISON
SMITH, JOSEPH HOWARD
SMITH, JOSIAH MORRISON
SMITH, SUMMERFIELD
STRICKLER, JAMES ALEXANDER
STRICKLER, WILLIAM L.
STUART, GEORGE WASHINGTON CUSTIS
SUBER, ALEXANDER
SWANN, ROBERT W.
SWISHER, BENJAMIN FRANKLIN
SWISHER, SAMUEL S.
TAYLOR, CHARLES SINCLAIR, JR.
TAYLOR, STEPHENS MASON
TAYLOR, WILLIAM M.
THOMAS W. R.
THOMPSON, LUCAS PLUST, JR.
TIDBALL, THOMAS ALLEN
TOMLINSON, AMBROSE
TOMPKINS, JOHN FULTON
TRENT, J. M.
TREVEY, JACOB MOSES
TRUEHEART, CHARLES WILLIAM
TYLER, JOHN ALEXANDER
UPTON, ROBERT A.
VEERS, CHARLES O.
WADE, THOMAS MORRALL, JR.
WALKER, JAMES STEEL
WALLACE, ALEXANDER AUGUSTUS
WATKINS, JOHN HASTINGS
WELSH, SAMUEL R.
WHITMORE, GEORGE W.
WILLIAMS, J. A.
WILLIAMSON, THOMAS W.
WILLSON, WILLIAM MATTHEW
WILSON, CALVIN
WILSON, H. A.
WILSON, JOHN A.
WILSON, SAMUEL ALEXANDER
WINSTON, ROBERT BACON
WISEMAN, WILLIAM BEARD
WOLFFE, BERNARD LIKENS
WRIGHT, JOHN W.
YOUNG, ROGER A.

NELSON, FRANCIS K., JR.
NELSON, PHILIP W.
NICELY, JAMES W.
O'ROURKE, FRANK
OWEN, THOMAS E.
PACKARD, WALTER JONES, Jr.
PAINE, CHARLES
PAINE, HENRY M.
PAINE, JAMES ALBERT
PASHCAL, R. B.
PAXTON, JACOB A.
PAXTON, SAMUEL A.
PAXTON, SAMUEL W.
PENDLETON, DUDLEY DIGGES
PHILLIPS, CHARLES CARTER
PLEASANTS, ROBERT A.
PORTER, MONIA GEORGE
POWELL, W.
PUGH, GEORGE W.
PUGH, LORENZA N.
RADER, HENRY
RAYNES, ARCHIBALD G.
RHODES, D. P.
ROBERTSON, JOHN W.
ROBINSON, WILLIAM F.
RUFFIN, JEFFERSON R.
SANFORD, JAMES III
SCHAMMANHONE, JOHN Z.
SHANER, JOSEPH FAUBER
SHAW, CAMPBELL A.
SHOULDER, JACOB MORTER
SINGLETON, WILLIAM FRANCIS
SMITH, JAMES POWER
SMITH, JOSEPH F.
SMITH, SAMUEL CUNNINGHAM
SPARROW, THOMAS G.
STRICKLER, JOHN JAMES
STRIDER, WILLIAM M.
STUART, WILLIAM CLARENCE
SWANN, MINOR W.
SWANN, WILLIAM MINOR
SWISHER, GEORGE WASHINGTON
TATE, JAMES FRAZIER
TAYLOR, ISAAC M.
TAYLOR, WILLIAM
THARP or THORPE, BENJAMIN F.
THOMPSON, JOHN A.
THOMPSON, SAMUEL GIVENS
TIMBERLAKE, FRANCIS H.
TOMLINSON, JAMES W.
TRAYNHAM, JOHN B.
TREVEY, DANIEL J.
TRICE, LEROY F.
TYLER, DAVID GARDINER
TYLER, RICHARD B.
VAN PELT, ROBERT
VEST, ANDREW J.
WALKER, GEORGE ALEXANDER
WALKER, JOHN WILLIAM
WALLACE, JOHN A.
WELCH, JOSEPH C.
WHITE, WILLIAM HOUSTON
WHITT, ALGERNON SIDNEY
WILLIAMS, JOHN JAMES
WILLIAMSON, WILLIAM GARNETT
WILSON, A.
WILSON, CHARLES A.
WILSON, JOHN
WILSON, JOHN ADJAR
WILSON, WILLIAM MONTGOMERY
WISEMAN, WILLIAM
WITHROW, JOHN E.
WOODY, HENRY
YOUNG, CHARLES EDWARD

ROCKBRIDGE DRAGOONS, COMPANY C, 1ST VIRGINIA CAVALRY

This company was organized as part of the militia at Fancy Hill on May 12, 1859. The Dragoons enlisted at Lexington on April 18, 1861.

Lieutenant Colonel CHARLES FRANCIS JORDAN

ADAIR, WILLIAM H.
ALEXANDER, JOHN McDOWELL
ARMENTROUT, CORNELIUS M.
ARMENTROUT, JOHN H.
AUSTIN, WILLIAM H.
BARCLAY, ELIHU H.
BARE, GEORGE
BARGER, WILLIAM PRESTON, JR.
BOWLIN, JOHN P.
BREWER, C.
BROCKENBROUGH, WILLIAM S.
BURKE, W.
BURKS, M.: Lt.
CAMERON, GEORGE HUGH
CAMERON, JOHN HYDE
CAMPBELL, JAMES M.
CHANDLER, NORBORNE EATON
CHANDLER, SAMUEL TEMPLE: Surgeon
CHITTUM, JAMES ALEXANDER
COMPTON, JAMES, Jr.
COOPER, ROBERT S.
CUMMINGS, FRANKLIN
DAVIDSON, GIVENS K. B.
DAVIDSON, ROBERT G.
DAVIS, JAMES
DILEN, S.
DIXON, JOHN JORDAN
DUNLAP, JOHN McKEE
EFFINGER, GEORGE WILLIAMS
ENTSMINGER, LEWIS G.
ERVINE, JOHN HAMILTON
FERRELL, MOSES
FIGGAT, JAMES SPENCER
FISHER, JOHN A.
FORD, DAVID HOUSTON
FULLER, SAMUEL B.
GHOLSON, W. A.
GILBERT, JOHN
GLOVER, ANDREW Y.
GOLD, JAMES McDOWELL
GOLD, SAMUEL, JR.
GORY, A. J.
GOUL, SAMUEL C.
GRAHAM, WILLIAM LYLE: Captain
GREENLEE, ELISHA GRIGSBY: Surgeon
GREENLEE, JOHN MARSHALL
GRIGSBY, LUCIAN PORTER
HAMILTON, WILLIAM WATTS
HANGER, JAMES R.
HANGER, MICHAEL R.
HARLAN, GEORGE BOYD
HARLAN, JERU SCOTT
HARLAN, WILKIE H.
HARRIS, JAMES F.
HATCHER, DAVID M.
HEMINGTON, W.
HOLDEN, JOHN S.
HOLDEN, THOMAS W.
HUMPHRIES, WILLIAM T.
JOHNSON, JAMES W.
JOHNSTON, WILLIAM J.
JORDAN, JOHN J.
KELLY, JOSEPH
LACKEY, HARVEY
LACKEY, ISAAC CARUTHERS
LACKEY, JAMES THOMAS
LACKEY, JOHN THOMAS

AGNOR, SAMUEL L.
ALLEN, DAVID
ARMENTROUT, HENRY
ARNOLD, HARRY J.
BALL, A. J. S.
BARE, ADAM
BARE, ISAAC
BARTON, ROBERT R.
BOWLIN, WILLIAM H.
BREWER, ISAAC
BUCKNER, E. PAXTON
BURKS, CHARLES R. P.: Lt.
CAMERON, CHARLES JACOB
CAMERON, GEORGE W.
CAMPBELL, CHARLES WILLIAM
CAMPBELL, WILLIAM C.
CHANDLER, S. T.
CHAPMAN, JAMES
CLEAVES, A. B.
CONNER, THOMAS BENTON
CRIGLER, DANIEL ARTHUR
CUMMINGS, JOHN S.: 1st Lt.
DAVIDSON, LEWIS C.: Lt.
DAVIDSON, WILLIAM D., JR.
DAVIS, SAMUEL
DIXON, GEORGE DOUGLASS
DUNLAP, JOHN MATTHEW
DUNN, THOMAS J.
ELHART, ADOLPH
ESTILL, ROBERT KYLE
EVANS, WILLIAM
FIGGAT, CHARLES MILES
FINKS, H. W.
FLOYD, ROBERT WILSON
FORD, ROBERT
GAITHER, GEORGE W.
GILBERT, EZEKIEL
GILMORE, ANDREW JACKSON
GOLD, JAMES A.
GOLD, JOHN W.
GOLD, SAMUEL McDOWELL
GOUL, JAMES P.
GRAHAM, EDWARD LACY
GREENLEE, DAVID ROBERT BARTON
GREENLEE, JAMES SAMUEL
GREENLEE, WILLIAM WOODVILLE
HAMILTON, JOHN GILBREATH
HANGER, AUGUSTUS T.
HANGER, JOHN B.
HARDEN, B. J.
HARLAN, WILLIAM HUNTER
HARLAN, SILAS G.
HARPER, CALVIN MOORE
HARTIGAN, WILLIAM PIPER
HATCHER, WILLIAM LOGWOOD
HILL, JOHN R.
HOLDEN, SAMUEL M.
HUMBLES, JAMES (Colored)
IMBODEN, JOHN ALEXANDER RUFF M.
JOHNSTON, JAMES MONTGOMERY
JOHNSTON, THOMAS C.
JORDAN, FRANCIS S.
KELLEY, JEREMIAH
KIRKPATRICK, JAMES L.
LACKEY, JAMES MORRISON
LACKEY, JAMES C.
LACKEY, JOHN H.

LACKEY, NATHAN HARVEY
LAIRD, HENRY RUFFNER
LAIRD, SAMUEL McKEE
LANTZ, F.
LAVERDAY, JOHN
LEAKE, ROBERT SHARPE
LEYBURN, ALFRED, JR.
LYNN, B. W.
MACKEY, SAMUEL C.
MARKS, WILLIAM HENRY
MARTIN, GEORGE
MEADE, WILLIAM T.
MILLER, BENJAMIN FRANKLIN
MONTGOMERY ANDREW SCOTT (2)
MONTGOMERY, JACKSON G.
MONTGOMERY, JOHN GILMORE
MOORE, HARRY ESTILL
MOORE, JOHN WILSON: Lt.
MOORE, SAMUEL RAMSEY
MORRISON, ROBERT CULTON
MYERS, HENRY H.
McCHESNEY, JAMES ZECHARIAH
McCLUNG, JAMES McDOWELL
McCORKLE, WILLIAM PHILANDER
McCROHIN, J.
McFADDIN, WILLIAM HENRY C.
McGREEVY, DENNIS
NICELY, HEZEKIAH
PARRY, JOHN McKEE
PATTON, JOHN A.
POAGUE, JAMES EDGAR: Ensign
POINDEXTER, WILLIAM BOWYER
PRESSY, CYRUS
PULTZ, JACOB
PULTZ, LEVI
RHODES, JACOB N.
ROBERTSON, JOHN WILLIAM
ROBINSON, JACOB A.
RUFF, JAMES WILSON
ST. CLAIR, WILLIAM C.
SCHINDELL, CHARLES F.
SEBERLING, A.
SEBERT, CHARLES D.
SMITH, BARTON
SUPINGER, JACOB A.
TALIAFERRO, PEACHY R.
TAYLOR, WILLIAM H.
THOMPSON, AUGUSTUS CHAPMAN, Jr.
TREVEY, CYRUS W.
TRIBBETT, WILLIAM W.
TURPIN, JAMES P.
UNROE, ADAM
WARREN, B. F.
WASH, WILLIAM JAMES
WEST, JOHN W.
WHITE, THOMAS SPOTTSWOOD
WILMORE, JACOB HENRY
WILSON, JAMES B.
WILSON, ROBERT K.
WINSTON, HENRY
WITHROW, JAMES McNUTT
WRIGHT, JOHN W.
ZOLLMAN, ADAM J.
ZOLLMAN, MADISON

LACKEY, NATHAN ALEXANDER
LAIRD, DAVID EDWARD
LAIRD, JOHN EWING
LAM, ANDREW CALVIN
LAVELL, ABRAHAM
LAVERDAY, WILLIAM
LEECH, THOMAS LACKEY
LUCAS, WILLIAM
MACKEY, JAMES PHILANDER
MALLORY, GILBERT ALEXANDER
MARTIN, ALEXANDER J.
MEADE, WILLIAM ZECHARIAH
MICHIE, CHARLES QUINICY
MONTGOMERY, ANDREW SCOTT (1)
MONTGOMERY, B. S.
MONTGOMERY, JOHN ALEXANDER
MONTGOMERY, THOMAS LACKEY
MOORE, JOHN W.
MOORE, RICHARD L.
MORRISON, HENRY RUFFNER
MORRISON, ROBERT HALL, SR.
MYERS, JOHN DANIEL
McCLINTIC, JOHN HENRY
McCORKLE, WILLIAM DOUGLASS
McCOWN, ROBERT McDOWELL
McFADDIN, ABRAHAM
McGOVERN, THOMAS W.
McNUTT, JAMES MORRISON
ORBISON, DAVID
PATTERSON, SAMUEL F.
PAULEY, THOMAS
POAGUE, JAMES WILSON
POWERS, OLIVER B.
PULTZ, CHARLES
PULTZ, JOHN A.
RADER, ZEBULON C.
RHODES, JOHN N.
ROBINSON, ANDREW D.
ROOT, IVERSON S.
RUFF, JOHN ANDREW
SALE, WILLIAM H.
SCOTT, THOMAS LACKEY
SHAFER, SAMUEL JACOB
STREET, JAMES POLK
SUPINGER, ROBERT
TAYLOR, ARCHIBALD M.
THOMPSON, ALEXANDER A.
TREVEY, ADAM SHANER
TREVEY, DAVID A.
TUNE, W. S.
TURPIN, NASH
WALKER, GEORGE S.
WATSON, W. A.
WELCH, WILLIAM LUCAS
WHITE, MATTHEW X., JR.: Captain
WILLSON, SAMUEL ALEXANDER
WILSON, ______
WILSON, JOSEPH S.
WILSON, SAMUEL L.
WIRT, JOHN
WITT, DAVID HENRY JR.
YEWSEN, H.
ZOLLMAN, JOHN WILLIAM

ROCKBRIDGE MEN IN OTHER COMPANIES OF THE FIRST VIRGINIA CAVALRY

BARTON, H. C.: Co. F
DAVIDSON, JOHN B.: Co. F
ESTILL, JOHN LIVINGSTON: Co. E
HENRY, ALEXANDER HORACE: Co. E
KOONTZ, ALEXANDER L.: Co. F
MASON, JAMES A., JR.: Co. B

BATIS, NORVAL WILLIS: Co. E
DAVIS, LEWIS PAYNE, JR.: Co. H
HARRIS, CHARLES H.: Co. E
KENNEDY, ISAAC: Co. E
LINCOLN, JACOB BROADDUS: Co. F
McLEOD, ROBERT TILDEN: Co. F

ROWAN, WILLIAM WALKER: Co. E
THOMAS, JACOB: Co. B
WHITE, EMORY ASBURY DULIN: Co. A
TURK, JAMES ALEXANDER: Co. E
WALKER, THOMAS JOHN: Co. E

SECOND ROCKBRIDGE ARTILLERY

The Second Rockbridge Artillery was organized at Fairfield on June 13, 1861 as the Fairfield McDowell Guards in honor of Miss Lillie McDowell, daughter of former Governor James McDowell. The complete history of this battery and its personnel is in Robert J. Driver's *The 1st and 2nd Rockbridge Artillery,* 1987.

ALEXANDER, WILLIAM POWHATAN
ALLEN, JAMES GEORGE
BARNETT, BENJAMIN FRANLKIN
BARTLEY, WILLIAM J.
BOWMAN, JOHN
CAMPBELL, JAMES A. J.
CAMPBELL, NIMROD McPHERSON
CAMPBELL, WILLIAM ALEXANDER, JR.
CASH, BRAXTON D.
CASH, JAMES W.
CASH, JOHN WASHINGTON
CASH, JOSEPH MARION
CASH, WILLIAM
CAVE, JOHN C.
CHANDLER, WILLIAM
CLEMMER, WILLIAM LEWIS
COCHRAN, ANDREW A.
COFFEY, PETER J.
COFFEY, RUBEN
CRIST, WILLIAM McCLUNG
CULTON, ZACHARIAH JOHNSTON
DECKER, HOWARD WESLEY
DONALD, WILLIAM KEYS, JR.: Captain
DRAIN, LEWIS C.
DRAWBOND, WILLIAM HENRY
DURHAM, EUGENE E.
FORD, JAMES PRESTON
FORD, WILLIAM ALEXANDER
GISINGER, JOHN TIMOTHY DWIGHT
GOOLSBY, WILLIAM CYRUS
GRIFFIN, LEWIS ANDREW
HAMILTON, HARVEY
HAMILTON, JOHN F.
HAYSLETT, ANDREW J.
HEIZER, EDWARD NEWTON
HINTY, WILLIAM HENRY
HITE, NATHANIEL WILSON
HITE, WILLIAM PAUL
HOUCHENS, W. MOSES PERRY
HOYLMAN, JOHN BRAINARD
HUGHES, JOHN PRESTON
HUMPHREYS, JAMES G.
JARVIS, JAMES EVERETT
JENKINS, CHURCHILL
KEFFER, HENRY
KERR, LORENZO DOW
LACKEY, THOMAS
LAWHORN, WILLIAM
LEECH, LUCIAN THEODORE
LONG, WILLIAM M.
LUCK, L. T.
LUSHBAUGH, WILLIAM L.
LYNN, JOHN CALVIN
MANN, JOHN A.
MILLER, DAVID LEWIS
MILLER, JOHN: Captain
MILLER, RICHARD S.
MOORE, ISAAC K.
MOORE, WILLIAM P.
ALLEN, DAVID W.
ALLEN, WILLIAM RICHARDSON
BARTLEY, HENRY A.
BEARD, HUGH S.
BRYANT, LORENZO SHAW
CAMPBELL, MATTHEW BRYANT
CAMPBELL, WILLIAM ALEXANDER, Sr.
CARVER, VALENTINE
CASH, JAMES PATTERSON
CASH, JOHN
CASH, JOHN WESLEY
CASH, JOSEPH W.
CASH, WILLIAM HENRY
CHANDLER, LINDSAY
CLEMMER, JAMES HARVEY
CLINE, DeWITT
COFFEY, MARVEL M.
COFFEY, ROBERT W.
COFFEY, WILLIAM MONTABALLE
CULTON, JAMES B.
CUPP, JOHN WESLEY
DICKINSON, JOHN CARTER
DOYLE, JOHN FLETCHER
DRAWBOND, JOHN LORENZA
DRAYTON, JOHN E.
EAKIN, JAMES MOORE
FORD, JOHN THOMAS
GAYLOR, WILLIAM
GOOLSBY, JAMES A.
GREEN, BENJAMIN F.
HAMILTON, GEORGE JULIAN
HAMILTON, HENRY W.
HAMILTON, WILLIAM LEWIS
HAYSLETT, ANDREW JACKSON: Surgeon
HESLEP, JOSEPH SPRIGGS
HITE, JOHN MARTIN
HITE, SAMUEL
HOLLER, JACOB B.
HOYLMAN, GEORGE W.
HUGHES, ELIJAH M.
HUGHES, WILLIAM CALVIN
HUMPHREYS, JAMES HENRY
JARVIS, JOHN
JOHNSON, ROBERT W.
KELLY, J. H.
KERR, WILLIAM D.
LAWHORN, PRESTON T.
LEECH, JAMES GRANVILLE
LILLEY, JAMES W.
LOVEGROVE, WILLIAM H.
LUDWICK, JOHN T. S.
LUSK, JOHN ANDREW MONTGOMERY: Capt
LYNN, WILLIAM T.
MEEKS, JOHN P.
MILLER, E. S.
MILLER, JOHN
MILLER, SAMUEL SMITH
MOORE, PATRICK
MORAN, NATHANIEL

MORRIS, DUDLEY M.
McCOWN, JOHN JAMES
McCRORY, WILLIAM THOMAS
McFADDEN, JOSEPH
McGUFFIN, WILLIAM W.
McNEIL, ROBERT J.
O'BRIAN, LAYFAYETTE
OTT, DAVID ALEXANDER
PATTERSON, JOHN M.
PATTERSON, WILLIAM DELANIE
PAUL, JAMES ANDREW JACKSON
PAXTON, JAMES HENDERSON
PAXTON, JAMES THOMAS
PAXTON, SAMUEL D.
PAXTON, WILLIAM A.
POTTER, SAMUEL MARTIN
PUGH, WILLIAM H.
RISK, JAMES PAXTON
ROBINSON, JAMES M.
SHEWEY, FRANKLIN
SHOVER, FRANKLIN
SLY, ADOLPHUS
SLY, JAMES DORSEY
SMILEY, WILLIAM ARCHIBALD
SORRRLLS, GEORGE D.
SORRELLS, THOMAS JEFFERSON
STRICKLER, ARCHIBALD W.
SWISHER, WILLIAM F.
TAYLOR, EDWARD ADOLPHUS
TAYLOR, JAMES EDWARD
TAYLOR, WILLIAM H.
THORN, JOB
TRIBBETT, ROBERT R.
VESS, MATTHEW
WALLACE, EDWIN, SR.
WALLACE, SAMUEL: 1st Lt.
WHITE, ISAAC M.
WHITE, ROBERT
WHITESELL, JEREMIAH
WHITESELL, ZACHARIAH TAYLOR
WILSON, JOHN A.
WILSON, SAMUEL WITHROW
WILSON, WILLIAM THOMAS: 1st Lt.
WINE, ROBERTSON ERWIN, SR.
WITHERS, CYRUS
WOOD, GEORGE GODFREY

McCORMICK, THOMAS NIMROD
McCRORY, EDWARD HUGHES
McDOWELL, THOMAS PRESTON
McGUFFIN, SAMUEL R.
McMANAMA, THOMAS PRESTON
McNUTT, ROBERT
ORENBURN, WILLIAM H.
PAINTER, JAMES HENRY
PATTERSON, WILLIAM A.
PATTERSON, WILLIAM L.
PAXTON, DANIEL JAMES: 3rd Lt.
PAXTON, JAMES POAGUE
PAXTON, JOHN W.
PAXTON, THOMAS N.
PEARL, JOHN
PUGH, JAMES H.
RAMSEY, DABNEY COLEMAN
RISK, JOHN W.
SELVEY, WILLIAM HARVEY
SHIELDS, WILLIAM CYRUS
SLOAN, JOHN COOKE
SLY, ALFRED A.
SMILEY, JOHN B.
SMITH, JOHN PARK
SORRELLS, JOHN JOSEPH
STEELE, JAMES E.
STUART, ERSKINE EBENEZER
TAYLOR, ARCHIBALD M.
TAYLOR, GEORGE WASHINGTON
TAYLOR, JOSEPH LARKIN
TEMPLETON, BENJAMIN FRANKLIN
TRIBBETT, FRANCIS MARION
VESS, CROSSBERRY DAVID
WALLACE, ALBRIGHT ALEXANDER
WALLACE, JOHN WILLIAM
WHITE, GEORGE W.
WHITE, JOHN P.
WHITESELL, JAMES W. C.
WHITESELL, JOHN WILLIAM
WILHELM, ABNER McCORKLE
WILSON, JOSEPH McC.
WILSON, THOMAS MITCHELL
WINE, JAMES A.
WISEMAN, ELIJAH MERCHANT
WOMELDORFF, JACOB B.

ROCKBRIDGE GRAYS, COMPANY H, 4TH VIRGINIA INFANTRY

This unit was organized in Lexington soon after the John Brown Raid, in 1859. The company reorganized for active service on April 21, 1861. See James I. Robertson's *4th Virginia Infantry*, 1982, for details on the history and the men who served. The roster has been supplemented from the author's files.

ACKERLEY, JAMES PAUL III
ACKERLEY, WILLIAM
AILSTOCK, THOMAS
ALEXANDER, ROBERT W. AUGUSTUS
ANDERSON, THOMAS A.
BARGER, DAVID W.
BARNETT, HENRY C.
BLACK, JOHN TUCKER
BOYTON, H. D. N.
BRYANT, ALEXANDER W.
BURCH, JOHN JAMESION
CAMDEN, JESSE
CAMDEN, WILLIAM F.
CARTER, JOHN
COX, JOHN D.
CUMMINS, ANDREW HAMILTON: 1st Lt.

ACKLERLY, JOHN S.
AILSTOCK, CHARLES PRESTON
ALEXANDER, ADOLPHUS A.
ANDERSON, JOHN B.
ARTHUR A.
BARGER, JOHN J.
BERRY, WILLIAM C.
BOWYER, JOHN T.
BROWNLEE, JOHN ARTEMUS
BUNCH, ANDREW J.
BURKS, CLIFTON CHARLES: 2nd Lt.
CAMDEN, OSCAR
CAMDEN, WILLIAM M., JR.
CLARK, ROBERT GILMORE
COX, SAMUEL
DENTON, WILLIAM

DONALD, JAMES C.
DONALD, ROBERT A.
DUDLEY, ROBERT
EADS, JAMES M.
EADS, ROBERT H.
EDMONDSON, DAVID TEMPLETON
ELLIOTT, JOHN McC.
ELLIOTT, SAMUEL P.
ENGLISH, M. M.
FAINTER, JAMES A.
FARROW, WILLIAM
FISHER, WILLIAM HENRY
FISHER, WILLIAM R.
FITZGERALD, JOHN F.
FORD, ANDREW D.
GARRETT, WILLIAM J.
GILES, EDWARD A.
GILMORE, HENRY CARTER
GOOLSBY, JOHN M.
HALL, JOHN W.
HALL, RICHARD
HAMILTON, ALEXANDER M.: Captain
HAMILTON, ANDREW JACKSON
HARRIS, PETER
HARRISON, JAMES T.
HARTSOOK, NEWTON B.
HAYES, BURRELL HILL
HEFFRON, EDWARD
HELMS, JAMES W.
HENDERSON, REUBEN D.
HENSLEY, JAMES D.
HENSLEY, JAMES G.
HENSLEY, JOHN G.
HICKS, JOSEPH
HILL, BENJAMIN F.
HILL, JAMES W.
HITE, BENJAMIN E.
HOGAN, PATRICK: 2nd Lt.
HYDE, THOMAS INGLIS
IMBODEN, HENRY DALLAS
IMBODEN, SAMUEL W. R.
JOHNSON, EDWARD
JOHNSTON, JAMES
KENNEDY, HENRY F.
LA BRIE, JOSEPH
LACKEY, WILLIAM H.
LAUCK, CHARLES EDWARD: 2nd Lt.
LAWRENCE, JACOB H.
LAWSON, JOHN E.
LECKEY, WILLIAM M.
LEECH, JAMES A.
LEECH, THOMAS HENRY
LEWIS, GEORGE W.
LEWIS, JOHN D.
LILLEY, JOHN A.
LILLEY, MARTIN MILLER
McCLURE, WILLIAM C.
McCORKLE, GEORGE BAXTER: Capt.
McCORKLE, JAMES T.
McCORKLE, WILLIAM ADAIR
McDANIEL, MATHEW WHITEMAN
McLANE, GEORGE W.
McMAMAMY, JAMES A.
McNABB, WILLIAM S.
MAJOR, JOSHUA BRIGHT
MARSTELLER, LE CLAIRE A.
MATEER, JOSEPH S.
MILLER, JOHN W.
MOFFETT, GEORGE HENRY
MOFFETT, JOHN STUART
MOFFETT, WILLIAM BARCLAY
MONTGOMERY, THOMAS LACKEY
MOORE, G. P.
MOORE, WILLIAM J.
MOORE, WILLIAM WARREN
MOXLEY, BENJAMIN P.
MULLEN, JAMES B.
NICELY, DUDLEY
NORTHERN, ROBINSON LUCAS
PATTON, WILLIAM M.
PAXTON, JOSEPH McCLUNG
PHILLIPS, JACOB
PHILLIPS, WILLIAM J.
PLEASANTS, JOSEPH J.
PRICHARD, JAMES F.
PRICHARD, JOHN
PUGH, JAMES H.
RAPP, BENJAMIN FRANKLIN
RAPP, SAMUEL G.
REYNOLDS, JAMES FRANK
REYNOLDS, LEWIS F. C.
RHODES, GEORGE PAYTON
RICKETTS, JAMES M.
RIGGS, W. H. C.
ROGERS, GRANDISON M.
ROGERS, WILLIAM H.
SHERVEY, WILLIAM
SILER, JOSEPH M.
SILVEY, JOHN F.
SILVEY, WILLIAM H.
SLOUGH, BAXTER
SLOUGH, BENJAMIN AUGUSTUS
SLOUGH, JAMES W.
SMALL, JOHN W.
SPENCE, JOSEPH
STERRETT, WILLIAM A.: 2nd Lt.
SULLIVAN, JOHN
TAYLOR, JOHN W.
THARP, HIRAM
THOMPSON, WILLIAM H.
TOMLINSON, MANSON
TOMLINSON, ROBERT
TRIBBETT, JOHN PERRY
TRIBBETT, JOSEPH F.
TRIBBETT, WILLIAM M.
TURNER, JOHN J.
TYREE, J. H.
UPDIKE, JAMES GLENN: Captain
VESS, PHILIP G.
VESS, THOMAS R.
WALLACE, ALEXANDER AUGUSTUS
WALLACE, WILLIAM P.
WEBB, JAMES HARVEY
WEBB, JOHN A.
WEBB, WILLIAM D.
WESTBROOK, HENRY A.
WILLS, ELISHA E.
WILMOUTH, WILLIAM A.
WILSON, JOHN BERRY
WILSON, JOHN McINTIRE
WILSON, SAMUEL P.
WILSON, THOMAS J.
WITHERS, JAMES E.
WITHROW, JOHN E.
WITT, WILLIAM EDWARD
ZOLLMAN, MADISON

COMPANY A, 4TH VIRGINIA INFANTRY

GLASGOW, ROBERT AUTHUR: 2nd Lt.

COMPANY G, 4TH VIRGINIA INFANTRY

WADE, WILLIAM: 1st Lt.

LIBERTY HALL VOLUNTEERS, COMPANY I, 4TH VIRGINIA INFANTRY

The Liberty Hall Volunteers was organized in Lexington on June 2, 1861. Fifty-seven of its original seventy-three members were students at Washington College. For the complete history of this company and its members, see James I. Robertson's *4th Virginia Infantry*, 1982. The roster has been supplemented from the author's files.

ADAIR, ALEXANDER
AMOLE, JAMES PRESTON
ANDERSON, JOHN REPOGLE, JR.
ANDERSON, ROBERT MILTON
ARNOLD, J. HARRY
BACON, ALGERNON SIDNEY
BAKER, SAMUEL DOLD
BARCLAY, JAMES W.
BEESON, HENRY M.
BELL, WILLIAM JAMES D.
BRIAN, J. A.
BROOKS, ANDREW
BROOKS, MOFFETT
BROWN, WILLIAM L.
BUCHANAN, JAMES M.
BUSHONG, ABRAHAM
CARR, R. N.
CASH, ROBERT C.
CHESTER, JOSEPH T.
CLYCLE, GEORGE A. E.
COPPER, JOHN M.
CRIST, GEORGE
DAVIDSON, GIVENS K.
DAY, WILLIAM E.
DUNLAP, ROBERT KERR
ECKARD, WILLIAM K.
GAYLOR, JAMES W.
GODWIN, THOMAS J.
GORDON, FRANKLIN
GROSE, JOHN
GWYNN, BRONSON B.
HALL, ANDREW H.
HARTLESS, ELI
HELMICK, WILLIAM HENRY
HODGE, JOHN S.
HUGHES, WILLIAM
HUTTON, GARDNER PAXTON
JACKSON, JOHN A.
JOHNSON, SAMUEL A.
JONES, JOHN HENRY BOSWELL: 2nd Lt.
KEENAN, JAMES
KAHLE, MATTHEW SALTZER
LACKEY, GEORGE WHITE
LACKEY, JOHN
LACKEY, THOMAS T.
LACKEY, WILLIAM N.
LAM, CYRUS MORRISON
LAREW, MILTON F.
LIGHTNER, SAMUEL MILTON
LINK, J. ABRAHAM TROXELL
LOGAN, NATHANIEL BURWELL
LONG, BRYON
LYLE, JOHN NEWTON: 1st Lt.
McALPHIN, ROBERT
McCLUNG, ANDREW ALEXANDER
McCOUGHTERY, JAMES W.
McCURDY, WILLIAM T.
ALMOND, REUBEN R.
AMOLE, THOMAS FRANKLIN
ANDERSON, JOSEPH MC.
ANDERSON, WILLIAM ALEXANDER
ARNOLD, JACOB WYATT
BAINE, G. W.
BARCLAY, ALEXANDER TEDFORD
BARTLEY, HENRY A.
BELL, CHARLES WILSON
BRADLEY, BENJAMIN A.
BROOKE, FRANCIS TALIFERRO
BROOKS, CHARLES
BROOKS, WILLIAM
BRYAN, DENNIS
BURKE, THOMAS NELSON
BYRD, JOHN A.
CASH, JAMES CLARK
CHAPIN, GEORGE W.
CLIFTON, ROBERT A.
COFFMAN, JOHN H.
CRIST, ABRHAM
CULTON, JAMES B.
DAY, SAMUEL M.
DUNLAP, JOHN MATTHEW
DUNLAP, SAMUEL McKEE
ERVINE, JOHN HAMILTON
GLASGOW, ALEXANDER McNUTT
GORDON, DAVIDSON J.
GREEN, THOMAS
GUY, JOHN R.
GWYNN, WORTH O.
HALLETT, ROBERT JAMES
HELMICK, DAVID
HENRY, W.
HOLT, JAMES W.
HUMPHRIES, W. S.
IRVINE, WILLIAM
JOHNSON, RICHARD J.
JOHNSTON, WILLIAM M.
JORDAN, HARRY E.
KERR, ROBERT M.
LACY, EDWARD A.
LACKEY, J. WILLIAM
LACKEY, NATHAN A.
LACKEY, WILLIAM HAMILTON
LAIRD, HENRY RUFFNER
LAMB, M. HENRY
LEWIS, EDWARD
LIGHTNER, JOHN P.
LOGAN, JOHN F.
LOGAN, SAMUEL B.
LUNSFORD, WILLIAM
LYLE, SAMUEL HARRISON: 1st Lt.
McCLELLAND, WILLIAM A.
McCLUNG, CHARLES BEALE, SR.
McCRAY, JOHN G.
McFADDEN, JOSEPH

McKEE, JOHN TELFORD
McVAY, MICHAEL
MEADE, EVERARD
MILEY, JOHN W.
MOORE, JAMES JULIUS
MOORE, JOHN POAGUE
MOORE, WILLIAM DORMAN
MORRISON, HENRY RUFFNER: Captain
MYERS, JOHN DANIEL
NELSON, CHARLES F.
OTT, HENRY ARTHUR
PAGE, COUPLAND RANDOLPH
PAXTON, ALEXANDER STUART
PAXTON, WILLIAM LINFIELD
PETTIGREW, JAMES MADISON
PRESTON, WILLIAM CARUTHERS
RAYMOND, JOSEPH SHERWOOD
REED, THOMAS W.
RIELY, JOHN WILLIAM
ROBERTS, JAMES
ROLLINS, THOMAS SEWELL
ROSEN, DAVID HARRISON
RUNDANELER, JOHN
SHECKEL, DAVID
SHIELDS, GEORGE WASHINGTON
SNIDER, A. J.
SNIDER, DANIEL
STEELE, WILLIAM D.
STONER, GEORGE H.
STRATTON, THOMAS C.
STRICKLER, GIVENS BROWN: Captain
TAYLOR, ISAAC M.
THOMPSON, WILLIAM JACKSON
TURNER, THOMAS MATTHEW
VARNER, JOHN ALEXANDER RUFF
WELSH, JOHN
WHITE, HUGH AUGUSTUS: Captain
WHITMORE, ANDREW J.
WHITMORE, E.
WILBOURN, WILLIAM ROBERT
WILHELM, LEVI
WILLIAMS, CHARLES
WILLSON, JOHN EDGAR
WILSON, JOHN THOMPSON
WITHERS, FRANCIS MARION
WOODS, JAMES WATSON
YOUNCE, WILLIAM

McNUTT, BENJAMIN F.
MACKEY, JAMES SAMUEL
MEADE, WILLIAM T.
MITCHELL, EDWARD A.
MOORE, JOHN F.
MOORE, SAMUEL RAMSEY
MOORE, WILLIAM W.
MYERS, HENRY H.
NEEL, CYRUS FRANKLIN
O'BRIEN, DENNIS
OTT, WILLIAM BAXTER
PATTERSON, WILLIAM LEDGERWOOD
PAXTON, HORACE ALEXANDER
PAYNE, JOHN C.
PETTIGREW, SAMUEL G.
RAMSEY, ALEXANDER B.
REDWOOD, JOHN TYLER
RICHARDSON, WILLIAM ESTIA
ROADCAP, DAVID LAIRD
ROBERTS, THOMAS H.
ROLLINS, WILLIAM H.
ROWSEY, LAFAYETTE F.
RUFF, DAVID EDMONDSON
SHERRARD, JOSEPH LYLE, JR.
SMILEY, WILLAM R.
SNIDER, ANDREW
SPOHR, JAMES W.
STERRETT, JOHN DOUGLAS
STONER, WILLIAM
STRICKLER, CYRUS DAVIDSON
SUDDARTH, JAMES LITTLETON
THOMAS, CHARLES D.
TREVEY, DAVID A.
UTZ, CALVIN
WATSON, JOHN GREENLEE
WESTBROOK, JAMES H.
WHITE, JAMES JONES: Captain
WHITMORE, DAVID
WHITMORE, GEORGE W.
WILHELM, A. M.
WILHELM, SAMUEL H.
WILLIAMS, SAMUEL P.
WILSON, HARVEY L., JR.
WILSON, WILLIAM NORVELL
WITHERS, MARION H.
YOUELL, WILLIAM HENRY

BOYS COMPANY, JUNIOR RESERVES, COMPANY F, 4TH BATTALION VIRGINIA RESERVES

This company was organized in Lexington on April 16, 1864 of boys 16-17 years old. Captain Charles William Freeman, only 17 years old, led the youths into battle at Piedmont on June 5, 1864. Following that battle the company fell back to Rockfish Gap and later served at Lynchburg during Hunter's raid. The company reorganized for the war on August 8, 1864 and served as Company F, 3rd (Christman's), Battalion Virginia Reserves to the end of the war. The company was mounted and served as guards at Libby Prison in Richmond and as couriers. Numerous members of this unit enlisted in other Virginia organizations during 1864-65.

BALDWIN, JOSEPH SALLING
BARGER, ROBERT A.
BENSON, WILLIAM ALEXANDER
BLACK, DAVID S.
BROWN, DAVID B.

BARGER, ALEXANDER
BELL, JAMES B.
BERRY, JAMES WILLIAM
BLUM, JOHN ALEXANDER: VMI Cadet
CAMPBELL, ISAAC NEWTON

CAMPBELL, JAMES M.
CASH, JOHN B.
CASH, MATTHEW BRYANT
CHAIRES, B. C.
COOPER, J. S.
COX, ELIJAH T.
DAILEY, JAMES
DAVIS, JOHN NEWTON
DIXON, JOHN
DRAIN, JAMES C.
DRISKELL, L. B.
ECHOLS, JOHN JORDAN
FIGGATT, ROBERT H.
FISHER, GEORGE W.
FORD, JAMES A.
FURR, WILLIAM M. G.
GOLD, JOHN W.
GOODBAR, CALVIN
GREY, LOGAN
HANGER, ANDREW SANFORD
HATCHER, EMMETT D.
HEPLER, SAMUEL M.
JESSUP, HENRY B.
JOHNSTON, THOMPSON P.
KIRKPATRICK, SAMUEL LINDSAY
LAM, ANDREW CALVIN
LEECH, ISAAC STEELE
LONG, AMOS
LUCAS, DAVID R.
MOHLER, WILLIAM A.
MOORE, J. S.
MORRIS, DAVID MILLER
McCLUNG, JOHN MOFFETT
McCOWN, ROBERT E.
McCULLOUGH, JAMES E.
McKINSEY, CUTHBERT B.
NEWLIN, EARL
PARSONS, JOHN STEELE
PATTON, HEZEKIAH McCLUNG
PAXTON, JOHN CALVIN
POAGUE, WILLIAM GORDON
PROSSER, RALPH HYLTON: VMI Cadet
RAMSEY, MILTON N.
REYNOLDS, WILLIAM CALVIN
ROBINSON, JOHN A.
SCHINDLE, CHARLES H.
SMITH, JOHN PERRY
SNIDER, HENRY G.
STRICKLER, WILLIAM
STUART, JOHN GERRARD
SWINK, GEORGE MILTON
TRIBBETT, JOHN FRANKLIN
VEST, SAMUEL G.
WATTS, PAULUS N.
WHITE, W. H.
WILSON, HORATIO THOMPSON
WILSON, JAMES ADJAR
WITHERS, GEORGE W.
WRIGHT, SCHUYLER BRADLEY

CAMPBELL, JOHN THOMAS
CASH, SCHUYLER BRADLEY
CHILDRESS, J. SAMUEL A.
COFFMAN, JOHN TAYLOR
COLONNIA, BENJAMIN AZARIAH: VMI Cadet
CUMMINGS, ROBERT ALFRED
DANIELS, W. C.
DE COST, JOHN
DIXON, SAMUEL
DRAWBOND, WILLIAM HENRY
DUNLAP, WILLIAM MADISON
ELLINGER, WILLIAM A.
FIREBAUGH, JOHN F.
FLEMING, ROBERT H.
FREEMAN, CHARLES WILLIAM: Captain
GARLAND, WILLIAM H.
GOLLADAY, J. M.
GOODBAR, JOHN M.
HAMILTON, ROBERT CLAY
HARDY, ALBERT
HENSLEY, GEORGE E.
HOLDEN, THOMAS W.
JOHNSON, SAMUEL McC.
KAHLE JACOB P.
LAIRD, ALEXANDER FRANKLIN: 2nd Lt.
LAM, J. CALVIN
LOGAN, CHARLES H.
LOWMAN, GEORGE W.
LYLE, DUNCAN CAMPBELL: 1st Lt.
MONTGOMERY, JOHN GILMORE
MOORE, SAMUEL P.
MORRIS, WILLIAM TAYLOR
McCORMICK, WILLIAM HENRY
McCOWN, SAMUEL THEODORE
McGUFFIN, THOMAS HARVEY
NEAL, JAMES C.
NUTTY, JOHN W.
PATTERSON, JOHN F.
PAXTON, ADOLPHUS A.
POAGUE, JOHN GORDON
PRING, JEFFERSON CALVIN
RAMSEY, C. J.
RANDAL, JOHN
ROBINSON, ALFRED G.
RUFF, ANDREW WALLACE
SHORT, JAMES MADISON MONROE
SMITH, ROBERT F.
STRICKLER, NATHAN L.
STRIDER, JOHN PHILIP
SWARTZ, JOSEPH GODFREY
THOMPSON, WILLIAM N.
TURK, JAMES ALEXANDER
VIA, WILLIAM M.
WATTS, THOMAS D.
WILHELM, ADAM
WILSON, JAMES A.
WILSON, JOHN WILLIAM
WRIGHT, SAMUEL

SOLDIERS FROM ROCKBRIDGE COUNTY WHO SERVED IN THE 5TH VIRGINIA INFANTRY

For information on this unit see Lee A. Wallace, Jr.'s *5th Virginia Infantry,* 1988.

AREHART, WILLIAM: Co. D
BEARD, DAVID W.: Co. D
BELL, CORNELIUS JACKSON: Co. D
BERRY, JAMES B.: Co. D

BARE, NEWTON MARION: Co. E
BEARD, WILLIAM SUMPTER: Co. D
BELL, JOHN CYRUS: Co. D
BRADLEY, JOHN HENRY: Co. E

BRADY, JAMES: Co. G
BUCHANAN, WILLIAM A.: Co. D
BUSH, JOHN WILLIAM: Co. C
CAMPBELL, ANDREW BAXTER: Co. E
CAMPBELL JAMES HAMLIN: Co. E
CAMPBELL, ROBERT J.: Co. E
CASH, JAMES M.: Co. E
CASH, JOSHUA: Co. E
CLEMMER, HENRY C.: Co. D
DAY WILLIAM HENRY HARRISON: Co. L
DUNLAP, WILLIAM H.: Co. E
FULTON, WILLIAM HARVEY: Co. D
GREAVER, ADAM A.: Co. E
HANGER, JACOB H.: Co. D
HARRIS, BLACKFORD G.: Co. H
HATTAN, JACOB: Co. E
HITE, PETER JACOBSON: Co. E
HODGES, JOHN W.: Co. F
HOUSER, JAMES W.: Co. E
HUMPHRIES, WILLIAM SHELTON: Co. E
JOHNSON, NOBLE T.: Co. A
MARKS, ROBERT S.: Co. C
MAYS, JOHN A.: Co. E
MOHLER, FREDERICK: Co. C
MORAN, CHARLES N.: Co. E
McCORMICK, ALFRED ADDERSON: Co. C
McCUTCHAN, JUDSON ODELL: Co. D
McKEMY, WILLIAM COOPER: Captain, Co. D
RAINES, WILLIAM J.: Co. H
REESE, GEORGE W.: Co. G
ROSEN, JOHN M.: Co. D
SPECK, SANFORD H.: Co. H
TAYLOR, JOHN CURTIS: Co. E
THORN, HENRY: Co. L
VIA, BENJAMIN F.: Co. E
WADE, JOHN B.: Co. D
WHEELER, EDWARD E.: Co. G
WILSON, WILLIAM H.: Co. G
WISEMAN, ROBERT H.: Co. D
BROOKS, JOHN W.: Co. E
BUCHANAN, WILLIAM WALKER: Co. I
BUTON, JOHN H.: Co. B
CAMPBELL, ANDREW THOMAS: Co. E
CAMPBELL, MATTHEW J.: Co. E
CARROLL, GEORGE FRANKLIN: Co. H
CASH, JOHN W.: Co. E
CASSIDY, WASHINGTON B.: Co. I
DAY, ANDREW M.: Co. L
DOYLE, LITTLE B.: 1st Lt., Co. G
EAST, WILLIAM M.: Co. L
GRAHAM, JOHN ALEXANDER: Surgeon
GREAVER, JOHN Y.: Co. L
HANGER, MICHAEL R.: Co. C
HARVEY, JAMES ALEXANDER: Co. E
HITE, HENRY SCOTT: Co. E
HITE, ROBERT STEELE: Co. E
HOLBERT, WILLIAM RICHARD: Co. E
HULL, JOHN McKEE: Co. C
INGRAM, JAMES M.: Co. C
LOTTS, SAMUEL L.: Co. D
MAYS, JAMES A.: Co. E
MILEY, ANDREW J.: Co. C
MONTGOMERY, THOMAS ALEXANDER: Co. E
MORAN, LEWIS ALEXANDER: Co. E
McCOWN, JAMES LARUE: Co. K
McKEMY, JOHN G.: Co. D
PAXTON, SAMUEL W.: Co. D
REED, JAMES FRANKLIN: Co. H
RHODES, JOHN JACOB: Co. K
ROSEN, THOMAS MARTIN: Co. D
SWEET, JACOB: Co. H
THOMPSON, JOSEPH H.: Co. E
TREVEY, JACOB MOSES: Captain, Co. C
WADDELL, EDWARD LIVINGSTON: 2nd Lt., Co. L
WEEKS, ELISHA B.: Co. E
WHISMAN, ISAAC: Co. F
WISE, JOHN HENRY: Co. I

ROCKBRIDGE SOLDIERS WHO SERVED IN THE 11TH VIRGINIA CAVALRY

See Richard L. Armstrong's *11th Virginia Cavalry*, 1989, for information on this unit.

AILSTOCK, CHARLES PRESTON: Co. G
AILSTOCK, ROBERT F.: Co. F
ANDERSON, SAMUEL: Co. F
ANDERSON, WILLIAM HENRY: Co. G
AREHART, ANDREW J.: Co. E
BOWYER, JOHN H. A.: Co. E
DODD, ROBERT DUNBAR: Co. E
EARHART, ANDREW J.: Co. A
GIBSON, JAMES: Co. F
GREENE, JAMES LANE: Co. E
HAMILTON, JOHN A.: Co. F
LARRICK, GEORGE BELL: Co. B
MARTIN, WILLIAM P.: Co. F
MORTER, JOHN LEWIS: Co. I
McCHESNEY, JAMES ZECHARIAH: Co. F
McDONALD, HARRY: Co. D
PEARCE, DANIEL E.: Co. A
SHERRARD, JOSEPH LYLE, JR.: Co. D
SMITH, JOHN T.: Co. E
SNEAD, WILLIAM JAMES: Co. F
WELSH, SAMUEL R.: Co. E
AILSTOCK, JORDAN DAVID: Co. G
AILSTOCK, ZOROBOREL: Co. G
AGNER, FERDINAND OSCAR: Co. E
BARTLEY, ANDREW J.: Co. K
BEARD, JOHN EDGAR: Co. E
BYRD, ROWLAND: Co. E
DOUGLASS, BURTON R.: Co. F
FRAZIER, JAMES ANDERSON: Co. F
GLENDY, BENJAMIN FRANKLIN: Co. G
GREEN, ZACHARIAH: Co. E
JAMES, SAMUEL: Co. E
LOWMAN, JAMES DAVIDSON: Co. G
MOFFETT, GEORGE HENRY: Co. unknown
McCHESNEY, ALEXANDER GALLATIN: Co. F
McCLUNG, CHARLES BEALE, JR.: Co. A
PALMER, JOSIAH H.: Co.'s I & K
SHERRARD, JOSEPH HOLMES, JR.: Lt., Co. F
SHORT, TILFORD: Co. A
SMITH, STUART T.: Co. G
WELSH, JOSEPH C.: Co. E
WITHROW, JOHN E.: Co. G

VALLEY REGULATORS, COMPANY K, 11TH VIRGINIA INFANTRY

This company was organized in the lower part of Rockbridge County and the northern part of Botetourt County on May 25, 1861. For details on this unit and the men who served see Robert T. Bell's *11th Virginia Infantry*, 1985. Some additional names have been added from the author's files.

AGNER, DAVID V.
AGNER, GEORGE WHITE
AGNER, WILLIAM CARUTHERS
AUSTIN, JOHN HENRY
AUSTIN, JOSEPH
AYERS, GEORGE W.
AYLOR, JOHN M.
BACH, SIGMOND
BASSELL, GEORGE WASHINGTON
BATES, FLEMING
BIBB, ROBERT
BLACK, LUCIAN JAMES
BOGGS, DANIEL A.
BOWYER, DAVID WILSON
BOWYER, JOHN D. H.
BRADFORD, PHILANDER SPOTSWOOD
BRAFFORD, HUGH WHITE
BRAFFORD, JOHN MARCELLUS
BROWN, JAMES NELSON
BROWN, RICHARD HENRY
BROWN, WILLIAM J.
CAMPBELL, JAMES WALKER
CAMPBELL, ROBERT C.: Lt.
CAMPBELL, SAMUEL L.
CAMPBELL, WILLIAM H.
CARR, JOHN
CASH, JAMES
CASH, JOHN
CASH, THOMAS F.
CHITTUM, THOMAS W.
CONNELLY, WILLIAM H.
CONNER, ______
COYLE, PETER
CRAWFORD, BILL DICK
CRAWFORD, WILLIAM S.
DIX, EDWARD THOMAS: Lt.
DOOLEY, ANDREW A.
DOOLEY, JAMES P.
DOSS, JAMES MADISON
DRISKELL, WILLIAM L.
DUPRIEST, F.
FALLS, BENJAMIN S.
FITZGERALD, JEREMIAH
FORTUNE, JOHN
FURGESON, ELDRIDGE
FURGESON, JAMES WASHINGTON
FURGESON, JOSEPH
FURGUSON, WILLIAM M.
GILMORE, THOMAS RUSSELL: Lt.
GORMAN, JOHN
GRADY, WILLIAM P.
HANNAH, GEORGE W.
HARDY, JAMES T.
HARDY, WILLIAM H.
HESLIP, TIMOTHY N.
HINES, B. W.
HOUSTON, ANDREW MATTHEW: Captain
HOUSTON, EDWARD MILLER
HOUSTON, THOMAS DIX: Captain
HOUSER, JAMES W.
HOUSER, JOSEPH H.
HUDSON, J. W.
HUDSON, ROBERT
HUDSON, THOMAS
HUDSON, WILLIAM
HUGHES, JOHNATHAN
HUNT, BARNARD B.
ISSACS, GEORGE T.
ISSACS, JOHN R.
JOHNSON, JAMES D.
JOHNSON, JOHN LUTHER
JOHNSON, ROBERT A.
JONES, JAMES WILLIAM
KEYTON, WASHINGTON J.
KIDD, JOHN W.
MARKHAM, JAMES W.
MARKHAM, VIRGAL A.
MARKHAM, WILLIAM T.
MARTIN, JAMES W.
MAYO, JAMES H.
MAYS, ISAAC S.
McCARTY, JOHN F.
McCLELLAND, ALFRED
McCLELLAND, GEORGE
McCLELLAND, JOSEPH E.
McCLELLAND, WILLIAM A.
McCULLOCH, ELI PETER
McCULLOCH, JOHN S.
McCULLOCH, JOSEPH C.
McCULLOCH, WILLIAM T.
MONTGOMERY, JOHN H.
MORRISON, JOHN M.
MORTON, WILLIAM A.
NEWCOM, WILLIAM H. H.
O'CONER, ______
OYLER, JOHN M.
PAINTER, JAMES J.
PAINTER, JAMES M.
PAINTER, JAMES N.
PAINTER, JOHN M.
PAINTER, MARTIN V.
PARKER, CALLAHILL
PARKER, HEZEKIAH THOMPSON
PARKER, RICHARD B.
PARKS, CHARLES SAMUEL
PAYTON, JOHN W.
POAGUE, JAMES WILLIAM RUTHERFORD
POWERS, JOHN
RAY, JAMES H.
RAY, JOHN A.
REED, STUART
REID, ANDREW SIEMON
REID, JAMES H.
REID, WILLIAM H. A.
REYNOLDS, JOHN J.
REYNOLDS, THOMAS E.
RHEA, JAMES H.
RICE, BENJAMIN K.
RICE, WILLIAM
RIPLEY, MADISON
SALE, GEORGE W.
SCHINDEL, JOHN H.
SCHINDEL, SAMUEL P.
SHEARER, H. C.
SHORTER, WILLIAM H.
SILER, JACOB A.
SILVEY, ALFRED
SILVEY, HARVEY
SILVEY, HENRY
SMITH, ANDREW JACKSON
TRENT, WILLIAM A.
UNROE, HENRY
WALKER, ELIJAH H.
WALKER, JAMES M.
WALKUP, JAMES DOUGLAS
WALKUP, SAMUEL HOUSTON
WALKUP, WILLIAM MADISON
WATKINS, JOHN K.
WILCHER, JAMES T.
WRIGHT, WILLIAM D.
WORTH, JOHN
YEATMAN, ALBERT ALLMOND: Captain

COMPANY A, 11TH VIRGINIA INFANTRY

BALLARD, JAMES F.
LAYNE, DAVID STAPLES
McCORKLE, WILLIAM PHILANDER
SLAGLE, JOHN H.
WALKUP, MATTHEW HENRY

COMPANY B, 11TH VIRGINIA INFANTRY

HARVEY, WILLIAM M.
POWERS, JAMES

COMPANY D, 11TH VIRGINIA INFANTRY

AMMEN, SAMUEL ZENUS
ANDERSON, CHARLES T.
HENKLE, JOHN M.
HAZLEWOOD, WILLIAM GREENVILLE
WILLIAMS, RICHARD BURKS
HOUSTON, DAVID GARDINER, JR.: Capt.

COMPANY G, 11TH VIRGINIA INFANTRY

GUY, DEWITT CLINTON
JOHNSON, GEORGE T.
SALE, JAMES POLK

VALLEY CAVALRY or RANGERS, COMPANY C, 14TH VIRGINIA CAVALRY

This company was organized at Churchville on May 15, 1862 from excess men from the Second Rockbridge Dragoons and the Churchville Cavalry and recruits. The majority of the men were from Rockbridge County and only those are listed below. For complete information on this unit and the men who served in it, see Robert J. Driver's *14th Virginia Cavalry*, 1988.

ACKERY, DAVID GARDINER
ACKERY, SHANKLIN M.
ALLEN, SAMUEL BROWN: 3rd Lt.
ALLEN, WILLIAM FRAISER: 3rd Lt.
BACHTELL, WILLIAM HENRY
BALTIMORE, JOHN H.
BRYAN, JAMES E.
CAMERON, ANDREW WARWICK, JR.: 3rd Lt.
CAMPBELL, ROBERT GRANVILLE
CAMPBELL, SAMUEL P.
CAMPBELL, WILLIAM
CLARK, JOHN E.
CLARK, SAMUEL C.
CLARK, WILLIAM
CUMMINS, JOHN F.
DRYDEN, PHILANDER EWING
FITZPATRICK, JOHN
FLINT, JAMES JOSEPH
FORD, WILLIAM A.
FRY, WILLIAM D.
GILMORE, ANDREW G.
GLOVER, ANDREW Y.
GOODBAR, JOHN B.
GREEN, SAMUEL R.
HATCHER, WILLIAM K.
HORN, JAMES
HORN, W. H.
HOUSTON, WILLIAM PAXTON
IRVINE, D. HUGH
IRVINE, JAMES A.
IRVINE, JOHN MONTGOMERY
IRVINE, MASLIN B.
IRVINE, THEODORE McCORKLE
KENNEDY, JOSEPH
KENNEDY, JOSEPH McC.
KIRKPATRICK, GIVENS McDOWELL
KIRKPATRICK, JAMES L.
KIRKPATRICK, JOHN ALEXANDER
KIRKPATRICK, SAMUEL LINDSAY
KNICK, ALEXANDER T.
KNICK, RICHARD M.
KNICK, SAMUEL G.
KNICK, WILLIAM
LACKEY, HORACE ARMSTRONG
LACKEY, JAMES WILLIAM
LACKEY, JOHN T.
LACKEY, THOMAS S.
LACKEY, THEOPHILUS SAUNDERS
LACKEY, WILLIAM ALEXANDER: Captain
LACKEY, WILLIAM HAMILTON
LAWHORN, JAMES MATHEW
MACKEY, JAMES SAMUEL
MACKEY, WILLIAM ALEXANDER
MARKS, GIDEON
McCUTCHAN, WILLIAM HAMILTON
McHENRY, SAMUEL ANDERSON
NICELY, CHARLES A.
NICELY, JOHN ALEXANDER
NICELY, MARION
RAMSEY, JOHN T.
REYNOLDS, JOHN A.
ROWLAND, ROBERT M.
SHAFER, ANDREW JACKSON
SHERIDAN, JOHN
SMITH, JAMES WILLIAM
SNIDER, ANDREW J.
SNIDER, JAMES T.
SNIDER, JAMES W.
SNIDER, JOHN DAVID
SNIDER, JOHN JACK

SNIDER, JOSEPH TREVEY
TEAFORD, DANIEL
TRIBBETT, MATHEW
VESS, JOHN A.
VESS, SAMUEL A.
VESS, WILLIAM H.
WILLSON, JAMES FRANKLIN
WILLSON, THOMAS MITCHELL
WILMORE, JAMES C. M.
WILSON, JOHN McCORKLE
TAYLOR, WILLIAM
TRIBBETT, JOHN FRANKLIN
TRIBBETT, WILLIAM
VESS, JOHN T.
VESS, WILLIAM ALEXANDER
WATERS, JOSIAH G.
WILLSON, JAMES HOWARD
WILMORE, ANDREW
WILSON, JAMES BROWN
ZOLLMAN, ALEXANDER MORRISON

Due to the popularity of the 14th Virginia Cavalry many recruits from Rockbridge County were assigned to fill up the ranks of the other companies in the regiment. Those identified are listed below:

AGNER, SAMUEL McD.: Co. G
ARMENTROUT, CHARLES H.: Co. G
AYRES, JOHN WILLIAM: Co. G
BARGER, JOHN J.: Co. D
BELL, ALEXANDER NELSON: Co. E
BROOKS, JOHN FINLEY: Co. F
BUNCH, WILLIAM H.: Co. G
DONALD, JAMES C.: Co. G
DUDLEY, THOMAS JOSEPH: Co. G
ERVINE, JOHN HAMILTON: No Co.
GOODLOE, JOSEPH BELL: Co. E
HEIZER, JAMES FRANCIS: Co. I
HENSLEY, JAMES G.: Co. G
HILEMAN, SAMUEL MARTIN: Co. B
JOHNSTON, JAMES: Co. G
LACKEY, JAMES HENRY: Co. D
MAJOR, JOSHUA BRIGHT: Co. G
MOFFETT, JOHN FLOYD: Co. D
MOORE, JAMES SCOTT: Co. G
NORTHERN, ROBINSON LUCAS: Co. G
PEARMAN, JOHN: Co. E
RAPP, BENJAMIN FRANKLIN: Co. G
SLOUGH, BENJAMIN AUGUSTUS: Co. G
SPECK, GEORGE M.: Co. I
SULLIVAN, JOHN: Co. G
VESS, PHILIP G.: Co. G
WHITMORE, DAVID: Co. I
ANDERSON, WILLIAM C.: Co. B
AUSTIN, ISAAC T.: Co. G
BARGER, CHARLES FRANKLIN: Co. D
BELL, ADAM DICKINSON: Co.'s A & K
BELL, JOSEPH C.: Co. E
BUNCH, JOHN JAMESON: Co. G
DICE, HENRY: Co. B
DOOLEY, T. CALEY: Co. G
EAGON, E. A.: Co. G
GLENDY, BENJAMIN FRANKLIN: Co. K
HARDY, GEORGE W.: Co. D
HELMS, MATTHEW E.: Co.'s B & I
HENSLEY, JOHN G.: Co. G
HUFFMAN, WILLIAM H.: Co. G
LACKEY, JOHN FRANKLIN: Co. D
LADY, JOHN BURFORD: Lt., No Co.
MILLER, JOHN W.: Co. G
MOFFETT, WILLIAM LEDGEWOOD: Co. D
MOORE, WILLIAM WARREN: Co. G
OCHILTREE, JAMES SAMUEL: Co. D
PIERSON, WILLIAM FRANKLIN: 2nd Lt., Co. L
ROWLINSON, JOHN: Co. D
SMALL, JOHN W.: Co. G
STONER, B. E.: Co. G
VESS, JACOB HENRY: Co. A
WEBB, WILLIAM D.: Co. G
WILLSON, ALEXANDER D.: Co. F

SECOND ROCKBRIDGE DRAGOONS, COMPANY H, 14TH VIRGINIA CAVALRY

The Second Rockbridge Dragoons was organized at Brownsburg in April, 1861. The men came from Brownsburg, Walker's and Hay's Creeks, Cedar Grove, Rockbridge Baths, Timber Ridge, Goshen and Bell's Valley sections of the county. For the complete history of this company and its men see Robert J. Driver's *14th Virginia Cavalry*, 1988.

Lieutenant Colonel JOHN ALEXANDER GIBSON

ADAMS, WILLIAM G.
ANDERSON, JAMES YOUELL
ANDERSON, JOHN Y.
ANDERSON, ROBERT BENJAMIN
ANDERSON, WILLIAM A. L.
BALSER, JOHN
BARTON, ROBERT THOMAS
BLACK, CALVIN LUTHER
BLACK, WILLIAM S.
BLAND, JOHN T.
BREEDLOVE, JAMES FRANCIS
BROWN, DAVID B.
BUCHANAN, CHARLES B.
ANDERSON, JACOB HORN
ANDERSON, JOHN A.
ANDERSON, JOHN YOUEL: 3rd Lt.
ANDERSON, SAMUEL B.
BAGLEY, HENRY WETHERELL, JR.
BALSER, SAMUEL
BEARD, WILLIAM LEWIS
BLACK, DAVID S.
BLACKWELL, WILLIAM M.
BLAND, WILLIAM JOHN
BROWN, ADAM McCHESNEY
BROWNLEE, WILLIAM KINEER
CALLAGHAN, JAMES W.

CALLAHAN, SALATHIEL M.
CAMPBELL, LAFAYETTE M.
CHITTUM, JOHN J.
CHITTUM, WESLEY TAYLOR
CHRISTIAN, P. H.
CONNER, JAMES C.
CULTON, JOSEPH A.
DAVIS, ARCHIBALD D.
DAVIS, LEWIS PAYNE
DICE, DAVID
DICE, JOHN
ELLINGER, ANDREW
FIREBAUGH, HENRY A.
FIREBAUGH, JAMES WILLIAM
FIX, WILLIAM H.
FORD, ANDREW TAYLOR
FOX, JAMES B.
FRIEND, ISAAC F.
GIBSON, JAMES SAMUEL
GILMORE, JAMES P.
GLENDY, JAMES LAREW
GREEN, CHARLES PRESTON
GREEN, JOHN WESLEY
GREINER, EDWARD GRANVILLE M.
GURRENT, C. M.
HALL, LORENZO DOW
HANGER, JOHN D.
HIGGINBOTHAM, REESE THOMPSON: 1st Lt.
HOUSTON, WILLIAM HALE
HOY, GEORGE C.
HUFFMAN, JAMES SOLOMON
HULL, NAPOLEON BONAPARTE
IRVING, CHARLES WILLIAM
JORDAN, GEORGE D.
KENNEDY, DAVID A.
KINNEAR, JOSEPH PATTIESON
LACKEY, WILLIAM ROBERT
LEECH, WILLIAM BOLIVAR FINLEY
LINDSAY, HORATIO THOMPSON
LOCKRIDGE, JAMES LANCELOT
LONG, FRANKLIN G.
LOWMAN, JOHN D.
LUDWICK, JACOB J.
LUNSFORD, ANDREW
LYLE, JOHN HART
LYNN, JOHN CALVIN
MACKEY, GEORGE A.
MACKEY, JOHN HENRY
MAHONE, JAMES C.
MILLER, ANDREW JACKSON
MOORE, ANDREW HARVEY
MOORE, SAMUEL H.
MOORE, WILLIAM WARREN (2)
MORTER, JOHN LEWIS
McBRIDE, JAMES JACKSON
McCHESNEY, JAMES ZECHARIAH
McCLUNG, ANDREW ALEXANDER
McCLUNG, DAVID BRAINARD
McCLUNG, JOHN TATE
McCLURE, NAPOLEON BONEPARTE
McCRAY, DAVID A.
McKEMY, JOHN H.
McNUTT, JOHN RICE: Captain
NEVINS, JAMES R.
NORCROSS, GEORGE W.
NORCROSS, WILLOUGHBY MABERRY
OTT, JAMES DAVIDSON
PALMER, CHRISTIAN
PARRENT, WILLIAM HENRY
PARSONS, NELSON
PATTERSON, HENRY HARRISON
PATTERSON, JOHN B.
PATTERSON, NIMROD HITE
PATTERSON, WILLIAM LEDGERWOOD
PAXTON, ABNER JERN

CAMPBELL, JOHN P.
CARWELL, JOHN H.
CHITTUM, THOMAS J.
CHITTUM, WILLIAM THOMAS
CLEMMER, JOSEPH ALEXANDER
CONNORS, OWEN
CULTON, ZECHARIAH JOHNSTON: 2nd Lt.
DAVIS, JAMES WILLIAM
DAVIS, WILLIAM W.
DICE, GEORGE WASHINGTON
DICE, WILLIAM
ERVIN, ANDREW M.
FIREBAUGH, JAMES BOLTON
FIREBAUGH, WILLIAM B.
FORD, ALEXANDER COOPER
FORD, WILLIAM BOWLES
FRANCIS, OWEN T.
FULWIDER, ROBERT D.
GILMORE, ARCHIBALD B.
GILMORE, THOMAS RUSSELL, JR.
GRANT, ALEXANDER D.
GREEN, HENRY A.
GREINER, CHRISTOPHER COLUMBUS
GREINER, JOHN H.
GUTHRIE, BENJAMIN
HANGER, JOHN BROWN, JR.
HERRIGAN, MICHAEL
HILL, LORENZO
HOUSTON, WILLIAM HOWARD
HUFFMAN, JAMES H.
HUFFMAN, JOHN J.
HUTCHESON, ROBERT STEELE, JR.
JOHNSON, JOHN
KELBER, CHARLES H.
KENNEDY, HUGH
KIRKPATRICK, JOHN McCOWN
LAIRD, SAMUEL McKEE
LILLY, JOHN A.
LINDSAY, JAMES A.: 1st Lt.
LOCKRIDGE, JAMES T.
LOTTS, ISAAC N.
LUDWICK, GEORGE C.
LUDWICK, JAMES K. P.
LYLE, JAMES ARCHIBALD: 2nd Lt.
LYLE, WILLIAM ALEXANDER
MACKEY, ALFRED BAXTER: 2nd Lt.
MACKEY, HENRY HUNT
MACKEY, JOHN WILSON
MILLER, ANANIAS J.
MOORE, ANDREW A.
MOORE, JOHN K.
MOORE, WILLIAM WARREN
MORRISON, HENRY RUTHERFORD
MYNES, SAMUEL S.
McCHESNEY, ALEXANDER GALLATIN
McCHESNEY, ROBERT A.: 1st Lt.
McCLUNG, BENJAMIN FRANKLIN
McCLUNG, JAMES ARCHIBALD
McCLURE, JOHN PILSON
McCLURE, W. B.
McCUTCHAN, WILLIAM H.
McMASTER, SAMUEL McCONNELL
McNUTT, JOSIAH T.
NEWTON, CHARLES
NORCROSS, THOMAS J.
OTT, FRANKLIN AUGUSTUS
OTT, JAMES W.
PARRENT, FRANCIS MARION
PARSONS, CORNELIUS H.
PATTERSON, CYRUS
PATTERSON, JAMES W.
PATTERSON, JOHN M.
PATTERSON, SAMUEL AUGUSTUS
PATTON, JAMES FRANKLIN
PAXTON, ALBERT GALLATIN McNUTT

PAXTON, JOHN A.
PAXTON, JOHN J.
PAXTON, WESLEY R.
PETTIGREW, SAMUEL G.
PETTIGREW, WILLIAM
PHARES, JOHN N.
PINKERTON, BRAINARD MELANCHTON
PINKERTON, WILLIAM B.
POAGUE, JAMES WILSON
PULTZ, DAVID A.
PULTZ, GEORGE FRANKLIN
RANDOL, ALEXANDER
REED, JOHN H.
REYNOLDS, J. T.
RHEA, SAMUEL T.
RUNNELS, JAMES L.
RUNNELS, WILLIAM DAVIDSON
SALE, PHILIP B.
SALE, ROBERT M.
SALE, WILLIAM M.
SANDRIDGE, WILLIAM A.
SAVILLE, JOSEPH C.
SENSABAUGH, THOMAS
SHANER, JACOB HAINER
SHEA, SAMUEL
SHEWEY, RISK A. C.
SHORT, SAMUEL W.
SMALLWOOD, WILLIAM A.
SMILEY, ANDREW
SMILEY, JAMES A.
SNIDER, JAMES C.
SNIDER, JAMES HENRY
SNIDER, JOHN NICODEMUS
SPITLER, GEORGE W.
STEELE, JOHN CHURCHMAN
STERRETT, ROBERT D.
STERRETT, SAMUEL WILLSON
STERRETT, WILLIAM MADISON: 2nd Lt.
STONER, DAVID HENRY
STONER, JOHN N.
STRAIN, JAMES ALEXANDER: Captain
STRAIN, SAMUEL PRESTON
STRICKLER, RICHARD
STUART, ALEXANDER S.
STUART, JOHN GERRRAD
STUART, SAMUEL W.
SWISHER, DANIEL
SWISHER, DANIEL Z. T.
SWISHER, JAMES F.
TAYLOR, ANDREW H.
TAYOR, ARCHIBALD M.
TAYLOR, JOSEPH LARKIN
TAYLOR, LARKIN
TAYLOR, WILLIAM
TEMPLETON, FIELDING HOUSTON
TERRILL, HENRY L.
TERRILL, JAMES RICHARD
THOMPSON, HORATIO HOWE
THOMPSON, JOHN ANDREW
TREVEY, JOHN JOSEPH
TREVEY, JOSEPH J.
VINCENT, JOHN T.
VINES, WILLIAM H.
WALKER, ALEXANDER STUART
WALKER, CYRUS W.
WALKER, D.
WALKER, SAMUEL A.
WALKER, WILLIAM A.
WALKER, ZECHARIAH JOHNSTON: 2nd Lt.
WEIR, ALEXANDER HAMILTON
WEIR, SAMUEL H.
WELCH, JOHN H.
WHEAT, JOHN WILLIAM
WHITE, MATTHEW X., JR.
WHITE, ROBERT H.
WHITLOCK, JOHN N.
WHITMORE, JOHN HENRY
WILLSON, GEORGE EDWARD
WILLSON, HUGH ROBERT
WILLSON, JAMES A.
WILLSON, JAMES A. J.
WILLSON, JAMES ALPHEUS
WILLSON, JOHN ALPHEUS
WILLSON, JOHN EDGAR
WILLSON, MATTHEW DOAK
WILLSON, SAMUEL NORVELL
WILLSON, THOMAS McC.
WILLSON, WILLIAM NORVELL: 3d Lt.
WITHERS, HENRY ADDISON
WITHERS, JAMES
WITHROW, HENRY ARCHIBALD
WRIGHT, JOHN R.
WRIGHT, WILLIAM N.

ROCKBRIDGE GUARDS, COMPANY H, 25TH VIRGINIA INFANTRY

This company was organized at Brownsburg on April 23, 1861. The men were primarily from the Walker's Creek District of the county. See the forthcoming volume, *25th Virginia Infantry,* for the history of this unit and the men in it.

ADAMS, HUGH
AGEE, B. G.
ALEXANDER, TREVELLA
ALLEN, LEROY
ANDERSON, JAMES DAVISON
ANDERSON, JOHN YOUEL
ANDERSON, SAMUEL
BALSER, BENJAMIN, JR.
BALSER, SAMUEL
BARE, JOHN
BARNES, DAVID J.
BEATY, HUGH A.
BEATY, WILLIAM (1)
BEATY, WILLIAM (2)
BENSON, JACOB
BENSON, JOHN G.
BENSON, JOHN W.
BENSON, PRESTON C.
BLACKWELL, MEREDITH C.
BLEDSOE, AUSTIN
BOSSERMAN, HENRY B.
BOSSERMAN, WILLIAM
BOSWELL, WILLIAM
BOWMAN, JOHN W.
BROWNLEE, HORATIO H.
BROWNLEE, ROBERT ALEXANDER
BROWNLEE, WILLIAM J.
BRYAN, JAMES E.
BUCHANAN, JOHN RICE: 2nd Lt.
BURTON, JESSE
CAMPBELL, JAMES G.
CAMPBELL, JOHN A.

CAMPBELL, ROBERT GRANVILLE
CLARK, JAMES YOUNG
CONNER, GEORGE
CURRY, DAVID P.: Captain
DEAVER, THEODORE A.
DECKER, WILLIAM C.
DICKEY, WILLIAM TELFORD
DIXON, ROBERT A.
DOOLEY, JACKSON H.
DUGGER, J. W.
FITCHETT, W. H.: Captain
FORBES, JASPER N.
GLENN, ISAIAH
GORDON, JAMES H.
GREEN, J. T.
HASKINS, NOAH H.: 1st Lt.
HITE, GEORGE W. (Little George)
HODGE, HENRY
JAMESON, M. W.
JARVIS, JOHN J.
JOHNSTON, J. M.
JONES, JOHN W.: Chaplain
KELLEY, SAMUEL M.
KENNEDY, MOSES W.
KERR, SAMUEL A.
KIMBERLING, JACOB
KIRKPATRICK, JOHN ALEXANDER
KIRKPATRICK, WILLIAM HENRY
LAIR, WILLIAM J.
LOWMAN, WILLIAM HENRY
LUCAS, DAVID R.
LUCAS, PETER
McCOWN, WILLIAM MONTGOMERY
McCURDY, ALFRED ALEXANDER
McCUTCHAN, JOHN FRANKLIN
McKEMY, JAMES LINDSAY: 2nd Lt.
McKEMY, WILLIAM DAVIDSON
McMULLEN, GEORGE
MACKEY, HENRY L.
MARKS, JAMES A.
MARTIN, WILLIAM P.
MATHENY, ARCHIBALD
MATHENY, JOHN H. (2)
MATTHEWS, JOSEPH
MAYO, JOSEPH J.
MILBY, C.
MONEYMAKER, DANIEL
MURRAY, JAMES
MYERS, ANDREW JACKSON
NICELY, JOHN ALEXANDER
NICELY, SALOME
NORCROSS, THOMAS J.
NUCKOLS, W. S.
PAINE, R. A.
PERRY, ERASMUS LAFAYETTE
PILLOW, J. H.
PRYOR, W. W.
RAMSEY, WILLIAM G.
REED, FRANKLIN S.
RICE, JOHN C.
RIPPETOE, ADAM
ROSEN, DAVID HARRISON
SCRUGGS, H.
SENSABAUGH, DAVID S.
SENSABAUGH, JOHN C.
SENTO, A. F.
SHERMAN, GEORGE F.
SHIPPLETT, JOHN M.
SNIDER, ABRAHAM
SNIDER, DAVID
SNIDER, JAMES CALVIN
SNIDER, JOHN JOE
STAGLE, F. F.
STRICKLER, DANIEL
STRICKLER, DAVID McC.

CHILDRESS, DAVID DILLARA
CLIFTON, JOHN M.
CRAVER, JOHN S.
DAY, CHRISTOPHER COLUMBUS
DECKER, SAMUEL H.
DENNIS, Y. W.
DIXON, JAMES WILLIAM
DIXON, WILLIAM
DRIVER, THOMAS H.
FIREBAUGH, ROBERT DUNLAP
FIX, JAMES
FORBES, SAMUEL A.
GODSEY, WILLIAM
GRAHAM, DAVID E.
HAMILTON, GEORGE N.
HILLERY, R. J.
HITE, GEORGE W.
HOOVER, HENRY LUTHER: 1st Lt.
JARVIS, JOHN G.
JARVIS, WILLIAM H.
JOHNSTON, N.
KELLEY, JOHN W.
KELLEY, WILLIAM H.
KERR, JAMES Mc D.
KERR, ZACHARIAH A.
KIRKPATRICK, CHARLES WATSON
KIRKPATRICK, ROBERT DAVIDSON
KOGER, J. D.
LOWMAN, JOHN FRANK
LUCAS, ANDREW
LUCAS, JOHN F.
LUCAS, PETER F.
McCOY, CHARLES DANIEL
McCUTCHAN, JAMES HARVEY
McCUTCHAN, JOSEPH H.
McKEMY, ROBERT ALFRED
McLAUGHLIN, JAMES M.
McQUAIN, HUGH
MAINWARING, GEORGE
MARTIN, A. J.
MASSIE, EDWIN BLACKWELL: 2nd Lt.
MATHENY, JOHN H. (1)
MATTHEWS, JEROME
MAUCK, JOSEPH
MILAM, JOHN N.
MONEYMAKER, ARCHIBALD B.
MONEYMAKER, JOHN C.
MUSIC, LEVI
MYERS, JACOB H.
NICELY, MARION
NOEL, WILLIAM G.
NUCKOLS, JOHN
OCHILTREE, JAMES SAMUEL
PATTERSON, JAMES A.
PERRY, WILLIAM WHITFIELD
PRICE, ROBERT H.
RAMSEY, JAMES A.
RAPP, WILLIAM A.
REID, JAMES W.
RIDLEY, BOB
ROBERTS, C. C.
ROSEN, WILLIAM TATE
SELBY, WILLIAM H.
SENSABAUGH, JACOB
SENSABAUGH, SAMUEL S.
SHERMAN, GEORGE BALLARD
SHERMAN, JOHN
SHOWALTER, HENRY POLK
SNIDER, DANIEL
SNIDER, JACOB STRICKLER
SNIDER, JOHN JACK
SNIDER, JOHN McC.
STOUT, A. S.: 1st Lt.
STRICKLER, DAVID
STRICKLER, JACOB C.

STRICKLER, JAMES
STUART, ROBERT A.: Captain
STUART, WILLIAM WALKER
SWEET, JOHN WALKER
'AYLOR, JAMES BALLARD
'EAFORD, JOHN H.
'HOMPSON, JOHN WESLEY
'RAINER, E. E.
VEST, JACOB HENRY
VIA, C. C.
WALKER, JOHN F.
WELLS, S. G.
WHITMORE, DAVID HENRY
WHITMORE, JOHN B.
WILLSON, JAMES HOWARD
WILLSON, THOMAS McC.
WILSON, JOHN THOMAS
WISEMAN, WILLIAM
WITHROW, JAMES W.
STRICKLER, SAMUEL
STUART, WILLIAM Mc C.: 2nd Lt.
SUDDARTH, FRANK
TAYLOR, HENRY HARRISON
TEAFORD, JACOB PAUL SEIG
THOMAS, WILLIAM WILSON
TOOMAN, WILLIAM HENRY ROGERS
VAN SANT, J.
VEST, PHILIP G.
VIA, WESLEY THOMPSON
WELCH, ROBERT ALEXANDER
WEST, ROBERT D.
WHITMORE, JACOB J.: Captain
WILLSON, JAMES ALPHEUS: Lt.
WILLSON, JOHN ALPHEUS
WILSON, JAMES ALPHEUS
WISEMAN, GEORGE WASHINGTON
WITHROW, HENRY JACOB VAN PELT
WRIGHT, GEORGE W.

COMPANY B

PATTERSON, BENJAMIN M.: 3rd Lt.

COMPANY C

McCORKLE, JAMES WILLIAM: 2nd Lt.

COMPANY D

LLISON, WILLIAM H.
LACKWELL, JAMES H.
BONDURANT, ALBERT L.

COMPANY G

LSTOCK, SIMON
SMITH, STUART T.

COMPANY I

OGSETT, WILLIAM R.
SEEBERT, LAUNCELOT L.

COMPANY K

PENNINGTON, SOLOMON

ROCKBRIDGE RIFLES, COMPANY H, 27TH VIRGINIA INFANTRY

This company was organized in November, 1859, as part of the ilitia. It entered state service as Company B, 5th Virginia Infantry in pril, 1861 and transferred to the 27th Virginia Infantry in July, 1861. For omplete information on this unit see the forthcoming volume on the *27th irginia Infantry.*

Brigadier General ELISHA FRANKLIN PAXTON

Colonel ANDREW JACKSON GRIGSBY

Lieutenant Colonel JAMES KERR EDMONDSON

'AMS, CHARLES A.
NOR, THOMAS BOWLEN
LEN, RICHARD H.
CHIBALD, JAMES
ERS, CHARLES M. G.
ADAMS, GEORGE W.
ALLEN, H.
ALLEN, ROBERT H.
AYERS, ALFRED G.
AYERS, MATTHEW P.

BAILEY, GEORGE WASHINGTON, JR.
BANE, ANDREW
BARTLETT, J. P.
BEARD, PETER
BEETON, ROBERT ELISON
BERRY, CHARLES G.
BOOGHER, EDWARD NICHOLAS
BOWYER, DAVID GUTHRIE
BREEDLOVE, JAMES WINCHESTER, JR.
BUMPAS, JAMES J.
CAMDEN, JOSEPH S.
CAMPBELL, HENRY T.
CAMPBELL, ROBERT HENRY
CARROLL, RUFUS
CHAPIN, GEORGE W.: 2nd Lt.
CHARLTON, JAMES JORDAN
CHARLTON, SAMUEL McCOWN
CHEEK, J.
CLAIRBORNE, JOHN
CONNER, DAVID Y.
CRAWFORD, HENRY
CRISER, JAMES H.
CUMMINS, JOHN ANDREW
DAVIDSON, PRESTON A.
DAVIS, MATTHEW
DEAVER, THOMAS HENRY
DODD, SHELTON F.
DONALD, JOHN A.
DRUMHELLER, DARIUS EDWARD
DRUMHELLER, THOMAS
EAST, JAMES W.
EDMONDSON, JOHN J.
ESKEW, WYATT J.
FISHER, HOWARD
FITZGERALD, PATRICK J.
FONSHILL, JACOB H.
GAITHER, G. W.
GORDON, NATHANIEL T.
GORDON, THOMAS
GREINER, ERASMUS
HALL, JOSEPH D.
HANEY, JUNIUS R.
HARRIS, JAMES F.
HARRISON, WILLIAM HENRY
HARTIGAN, JOHN WESLEY
HARTIGAN, WALTER DOUGLAS
HAWK, JOHN
HEILBONER, HENRY
HIGGINS, JOSEPH A.
HILEMAN, JOHN JOSEPH
HITE, ELEB F.
HITE, WILLIAM A.
HOPKINS, DAVID LAWRENCE
HOUSTON, WILLIAM HALE
HOYLMAN, JOHN
JESSUP, EDWARD THOMAS
JONES, HIRAM S.
KAHLE, WILLIAM HENRY HARRISON
KEATLEY, JAMES L.
KELLEY, THOMAS B.
KIRKPATRICK, THOMAS M.
KREMER, JOHN
LAIRD, DAVID EDWARD
LETCHER, SAMUEL HOUSTON: Captain
LOKER, WILLIAM H.
MARKS, DAVID
MAY, NATHADEM
McCAMPBELL, SAMUEL JOHN NELSON
McCLURER, ASBURY C.
McCOWN, ROBERT
McCULLOCH, SAMUEL HOUSTON
McNAMARA, LAWRENCE
MILLER, ADAM W.
MOODY, JOHN E.

BAKER, ______
BARGER, JOHN ALEXANDER
BAUPIT, JESSE
BEETON, JOHN HENRY
BELL, ROBERT B.
BONES, THADELEO BRECKINRIDGE
BOUDE, JOHN CLINTON: Captain
BOYD, ______
BREEDLOVE, JOHN W.
CAMDEN, JAMES B.
CAMDEN, WILLIAM P.
CAMPBELL, JAMES M.
CAMPBELL, SAMUEL J.
CHAMPE, ANDREW J.
CHAPIN, JOSEPH W.: Lt.
CHARLTON, JOHN ALEXANDER
CHARLTON, WILLIAM CORBIN
CLAIRBORNE, ALFRED
CLOWES, PAUL J.
CRAIG, WILLIAM A.
CRISER, CHARLES
CRIST, HENRY
DAVIDSON, FREDERICK
DAVIS, DAVID
DAWSON, W. A.: Lt.
DICKENSON, JOHN CARTER
DONALD, BENJAMIN M.
DRAIN, ROBERT
DRUMHELLER, JOHN LEONARD
DRUMHELLER, WILLIAM P.
EAST, JOHN J.
EDMONDSON, JOHN M.
EVANS, JAMES S.
FITCH, CHARLES H.
FLOYD, JAMES MARION
FULLER, SAMUEL B.
GILLOCK, JAMES WILLIAM: 1st Lt.
GORDON, SAMUEL A.
GREEN, WILLIAM L. T.
HALK, ALEXANDER G.
HANEY, JAMES R.
HANGER, MICHAEL R.
HARRISON, WASHINGTON H.
HARTIGAN, A. G.
HARTIGAN, ROBERT MARCHELLE
HARTMAN, HENRY THEOPHILUS
HAYSLETT, ANDREW
HICKMAN, JACOB
HILEMAN, DANIEL JACOB
HILL, HARRISON
HITE, JOHN
HOOK, NATHAN D.
HOPKINS, HENRY ST. GEORGE: Surgeon
HOYLMAN, DANIEL
HUTCHESON, JAMES HAMILTON
JOHNSON, RICHARD
JORDAN, RICHARD A.
KARR, ZACHARIAH A.
KELLEY, JEREMIAH
KELLY, JOHN W.
KREMER, CHARLES
LACY, WILLIAM A.
LEECH, ANDREW JACKSON
LEWIS, WILLIAM W.: 2nd Lt.
LYNCH, EUGENE
MARKS, WYATT LEWIS
McALEER, ROBERT EMMETT
McCLURER, ARTHUR D.
McCLURER, N. D.
McCOWN, ROBERT McDOWELL
McCURDY, ALFRED ALEXANDER
MIDDLETON, JOHN WILLIAM
MITCHELL, WILLIAM
MOODY, WILLIAM T.

MOORE, JOHN E.
MORGAN, E. H.
MULLEN, JOHN L.
NEFF, JOEL L.
PAINE, ROBERT
PARRENT, FRANCIS MARION
PATTERSON, JOHN HOWARD
PATTON, WILLIAM
PAXTON, JAMES T.
RADFORD, CALVIN
REILLY, C. A.
ROBERTSON, LEVI
ROLLINS, CHARLES A.
RUFF, JAMES WILSON
SEAL, MORGAN F.
SHELTMAN, MATHEW R.
SHIELDS, JAMES H.
SHIELDS, WILLIAM THOMPSON
SHOEMAKER, NICHOLAS N.
SIDERS, WILLIAM
SLOUGH, JAMES McDOWELL
SMITH, ALPHONSO
SMITH, JACOB
SMITH, JAMES W.
SPEARS, WILLIAM J.
STEELE, THOMAS
STONE, JOHN D.
TANQUARY, ALFRED B.
TEED, MOSES E.
THOMAS, TUCK
TOBIN, J.
TRIBBETT, ANDREW DAVID
VAN PELT, JESSE
VARNER, CHARLES VAN BUREN
WADE, ALGERNON SIDNEY
WALLACE, HORACE HOLMAN
WALLACE, WILLIAM
WASH, THOMAS H.
WHEAT, JOHN WILLIAM
WILSON, THOMAS J.
WINES, WILLIAM M.
WRIGHT, WILLIAM G.
MOORE, JOHN HARVEY
MULLEN, ANDREW JACKSON
MULLEN, THOMAS BRADLEY
NORGROVE, EDWARD W.
PARKS, JOSHUA, JR.
PATTERSON, ALEXANDER McD.
PATTERSON, WILLIAM
PATTON, WILLIAM McFARLANE
PLEASANTS, JOHN THOMAS
REID, GEORGE W.
REILLY, DANIEL McGUIRE
ROBINSON, WILLIAM M.
RUDDLE, CALVIN
RUFF, SAMUEL WALLACE
SENSENY, JAMES MADISON
SHELTON, M. H.
SHIELDS, JEFFERSON (Colored)
SHIELDS, WILLIAM W.
SHUPE, H. P.
SIZER, CHARLES W.
SLUSHER, FLOYD
SMITH, HENRY D.
SMITH, JAMES STOKELEY
SNIDER, JOHN
STANDOFF, HENRY H.
STEELE, WILLIAM D.
STRONG, CYRUS E.
TANQUARY, J. S.
THOMAS, JAMES S.
THOMPSON, JOHN S.
TRANSOM, WILLIAM D.
TURPIN, RICHARD C.
VARNER, ANDREW WALLACE: 2nd Lt.
VINES, ANDREW JACKSON
WADE, JACOB B.
WALLACE, JAMES W.
WALZ, WILLIAM
WEBB, PHILLP M.
WILHELM, JOHN W.
WINE, ROBERT E., SR.
WOODY, JOHN D.

COMPANY B, 27TH VIRGINIA INFANTRY

BEARD, JOSEPH C.
BENSON, HENRY H.
BIBB, JOHN H.
CHITTUM, WILLIAM H.
COX, JAMES M.
DAVIS, THOMAS C.
GAYLOR, JAMES T.
GORDON, WILLIAM L.
HENKLE, PRESTON D.
HIGGINS, JAMES LEM
HITE, WILLIAM S.
INGRAM, WILLIAM A.
LITTLE, WILLIAM A.
MILLER, W. D.
MUTERSPAUGH, JAMES F.
PAINTER, CHARLES PERRY
POWERS, WILLIAM G.
REYNOLDS, JOHN A.
SNIDER, WILLIAM B.
SWEET, JOSEPH
WELCH, JOHN K.
WOODS, JAMES
BEARD, WILLIAM
BENSON, JAMES FRANKLIN
BLACKWELL, WILLIAM M.
CHITTUM, WILLIAM WESLEY
DAVIS, CHALKLEY
FIX, ROBERT JACKSON
GAYLOR, JOSEPH E.
HART, JOHN HARVEY
HENRY, SCOTT
HILEMAN, PHILIP CHRISTOPHER
HORN, JOHN H.
LITTLE, JAMES H.
MASON, JAMES A., JR.
MILLER, WILLIAM E.
McCRAY, JAMES PERRY
PLOGGER, BLAINE A. A.
QUIGLEY, JOHN
SMITH, ALEXANDER F.
SPROUSE, WILLIAM H.
TEAFORD, DAVID
WELCH, JOHN W. P.: Captain

COMPANY C, 27TH VIRGINIA INFANTRY

BEESON, ISAAC J.
BROWN, JOHN
CUMMINGS, JOSEPH J.
BRADLEY, SAMUEL R.
CLARK, WILLIAM A.
CUMMINGS, SAMUEL A.

DONALD, ALEXANDER
ECHARD, JOHN H.
FORBES, JAMES H.
GAYLOR, JAMES T.
GRANT, JOHN M.
GREEN, HENRY
GREEN, WILLIAM H.
GROAH, MICHAEL
HALL, JOSEPH H.
HANGER, GEORGE M.
HAYSLETT, ANDREW
HAYSLETT, JOHN
HEMP, JOHN P.
HEMP, WILLIAM S.
HILL, THOMAS JEFFERSON
HITE, HENRY SCOTT
HOLLAR, ROBERT
HUFFMAN, PHILIP IRA
ICENHOWER, WILLIAM R.
INGRAM, WILLIAM A.
LINK, JAMES C.
MILEY, JOHN W.
MOORE, JOHN NEVIUS
MOORE, SAMUEL D.
POAGUE, JAMES E.
POAGUE, SAMUEL WATSON
POAGUE, WILLIAM LEMUEL
PULTZ, CHARLES
PULTZ, DAVID
RAMSEY, HENRY KERR
ROADCAP, CHRISTIAN
WHITMORE, ANDREW
WHITMORE, DAVID

COMPANY D, 27TH VIRGINIA INFANTRY

BALSER, BENJAMIN F.
BENSON, JOHN
DAVIS, CAMPBELL R.
DICE, WILLIAM
FORBES, J. H.
GAYLOR, JOHN
HALL, JOHN
HOTINGER, ABRAHAM
HOTINGER, ALLEN
HOSTETTER, ANDREW
LACKEY, JOHN FRANKLIN
LOTTS, DAVID L.
LUCAS, PETER
MILEY, JACOB
MILLER, ALFRED LORIN
MILLER, GEORGE W.
MOHLER, JOHN W.
MONEYMAKER, JOSEPH
McCRAY, THOMAS H.
NICELY, CHARLES A.
NUTTY, JAMES F.
PARKER, JAMES
PARKER, ROBERT M.
REYNOLDS, WILLIAM
SMILEY, HUGH
SMILEY, JACOB
SNIDER, ANDREW
SWISHER, HENRY
SWOOPE, HENRY
THOMPSON, MICHAEL A.
TOLLEY, WILLIAM P.
WILHELM, JAMES P.
WILHELM, JOHN H.
WILHELM, LEVI
WILSON, BENJAMIN W.
WILSON, FRANKLIN CARUTHERS: Capt.

COMPANY E, 27TH VIRGINIA INFANTRY

BIBB, JAMES H.
BLACK, JAMES L.
BLACK, RICE
BLACK, WILLIAM L.
BYRD, JOHN J.
CASH, MATTHEW BRYANT, SR.
CHRISTIAN, JAMES P.
CLARK, SAMUEL P.
CLARK, WILLIAM C.
CONNER, FITZALLEN Y.
COOLEY, JACOB F.
CUMMINGS, FRANKLIN
CUMMINGS, JAMES JOSEPH
CUMMINGS, ROBERT W.
EDWARDS, J. J.
GARRETT, JAMES H.
GOODBAR, GEORGE
GOODBAR, HARVEY
GOODBAR, SAMUEL
GREGORY, WILLIAM
GREEN, WILLIAM L.
HALL, JAMES C.
HALL, WILLIAM
HARRISON, HOWARD MILTON
HAYSLETT, GEORGE
HAYSLETT, ROBERT
HULL, WILLIAM M.
IRVINE, HUGH W.
IRVINE, WILLIAM E.
JOHNSON, GEORGE W.
JOHNSON, J. HENRY
JOHNSTON, CHAPMAN
LEECH, DAVID
MACKLIN, JAMES J.
MILLER, SAMUEL H. H.
MOORE, JOHN P.
MOORE, SAMUEL D.
MYERS, ALLEN
McFADDEN, ABRAHAM
REID, JOHN AMOS TAYLOR
REID, SAMUEL NEWTON
REYNOLDS, OBEDIAH B.
ROBERTSON, JOHN H. H.
ROBINSON, WILLIAM M.
SALLING, GEORGE JACKSON
SANDFORD, HENDERSON
SHAFER, SAMUEL JAHOD
SHORT, TILFORD
SHOWALTER, SAMUEL
SLOUGH, ABRAHAM ALEXANDER PRESTON
SORRELS, THOMAS JEFFERSON
WATTS, BAXTER T.
WATTS, LUTHER MADISON
WEBB, CHARLES WILLIAM
WEST, THOMAS IRA, JR.
WHITE, JAMES M.
WHITE, JOHN B.
WILSON, JAMES W.
WILSON, JOHN B.
WILSON, JOHN McCORKLE
WINE, WILLIAM M.

COMPANY F, 27TH VIRGINIA INFANTRY

ACKERLY, WILLIAM A.
ANDERSON, WILLIAM S. or C.
BALLARD, ANDREW JACKSON
BARGER, GEORGE W.
BLACK, GARLAND C.
BLACK, GEORGE C.
BOGGESS, BENJAMIN B.
BROWN, HENRY D.
BYRD, JOHN J.
CALLISON, SAMUEL W.
CARTER, JAMES MADISON
CASH, THOMAS F.
COOLEY, JOEL F.
CRANE, JOSEPH S.
DANIEL, ROBERT H.
DAVIS, CHARLES L.
DAVIDSON, JAMES G.
DOWNS, GEORGE
DULANEY, RICHARD A.
EWING, JAMES A.
EWING, JOHN McKENRY
FAINTER, WILLIAM F.
GRIMES, WYATT C.
HALL, ANDREW D.
HALL, JOSHUA
HALL, JOSHUA J.
HALL, THOMAS
HALLOWAY, A. J.
HANGER, ZENUS FREEMAN
HARTIGAN, JAMES A.: 2nd Lt.
HAYSLETT, BRADLEY
HESLIP, THOMAS H.
HOLT, THOMAS W.
JOHNSON, GEORGE W. (1)
JOHNSON, GEORGE W. (2)
JOHNSON, JOHN HENRY
JOHNSON, LAFAYETTE M.
JOHNSON, REUBEN A.
LAWHORN, JOSEPH M.
LEECH, WILLIAM ADDISON
MARTIN, ANDREW M.
MAYS, JOHN JAMES
MOORE, ANDREW
MOORE, HORACE C.
MOORE, ROBERT M.
McCLUNG, ROBERT A.
McCLURE, WILLIAM D.
McDANIEL, MATTHEW W.
McGRADY, JAMES W.
NUCKHOLS, GEORGE W.
PRING, ROBERT T.
PURSELY, WILLIAM
REID, A. WESLEY
REYNOLDS, GEORGE
ROWLINSON, JOHN D.
SAVILLE, SAMUEL
SILER, PHILIP MADISON
SMITH, SAMUEL S.
SWEENEY, LEROY
THOMAS, JOHN H.
WATKINS, GEORGE WILLIAM
WATKINS, WILLIAM G.
WEBB, JAMES MILLER
WYATT, REUBEN

COMPANY G, 27TH VIRGINIA INFANTRY

ACKERLY, JAMES M.
ACKERLY, JOHN S.
ALLEN, ALPHONSO SAMUEL
ALMOND, WILLIAM C.
BENSON, ISAAC J.
BLACK, RICHARD TANKERSLEY: Lt.
CAMPBELL, JAMES S.
CAMPBELL, WILLIAM ADDISON
COPPER, BENJAMIN F.
CUMMINGS, J. A.
DAVIS, WILLIAM J.
DECKER, GEORGE TATE
FENTON, STEVEN G.
FRAZIER, JAMES W.
GRAHAM, JAMES F.
HANGER, GEORGE S.
HELMICK, JOHN
HODGE, JOHN S.
HUFFMAN, PHILIP IRA
KETTLEWELL, SAMUEL H.
LA BRIE, JOSEPH
LADY, JOHN BUFORD: 2nd Lt.
LAMB, JOHN G.
LINCOLN,JACOB BROADDUS
LOCKRIDGE, EDWARD FRANKLIN
LUDWICK, HENRY A.
LUDWICK, SAMUEL JOSEPH
McCABE, JOHN S.
MILLER, THOMAS J.
MOYERS, E. FRANK
MYERS, EVAN F.
ORENBAUM, JOHN LEWIS
PATTON, ALBERT
PAXTON, JAMES T.
REED, SAMUEL F.
REESE, JAMES GRANVILLE
ROBERTSON, JAMES M.
ROBINSON, WILLIAM THOMAS
SHEWEY, JOHN L.
SHIELDS, JOHN B.
SHUEY, JOHN P.
SMILEY, ARCHIBALD
SNIDER, WILLIAM FRANKLIN
SWEET, JOHN P.
WALTON, THOMAS
WEEKS, ADAM
WOMELDORF, DANIEL B.
WOMELDORF, JOHN A.
WRIGHT, GEORGE W.

27TH VIRGINIA INFANTRY, COMPANY UNKNOWN

BLACKWELL, WILLIAM B.
CUMMINGS, ANDREW A.
DIGGS, WILLIAM J.
FORD, WILLIAM A.
FRY, JOHN J.
WILSON, JOHN M.

52ND VIRGINIA INFANTRY

Lieutenant Colonel JOHN DeHART ROSS
Lieutenant Colonel THOMAS H. WATKINS
Surgeon LIVINGSTON WADDELL

Captain THOMAS H. WATKINS, COMPANY E

This company was raised in the Colliers Creek, Buffalo Creek, Broad Creek and Natural Bridge areas of the county. The company enlisted in Confederate service at Staunton August 1, 1861. See Robert J. Driver's *52nd Virginia Infantry*, 2nd edition, 1989, for complete information.

ACKERLY, GEORGE W.
ARTHUR, JOSEPH D.
AYRES, ALFRED GRAHAM
BANE, ABNER Mc C.
BEARD, WILLIAM B.
BLACK, ANDREW H.
BLACK, JOHN
BRADSHAW, ALEXANDER K.
BURGER, JOEL
BYERS, WILLIAM CONRAD
CAMDEN, JAMES
CAMDEN, JOHN
CAMDEN, LAYNE or LANE
CASH, JOSEPH BENJAMIN
CLARK, JOSEPH DAVID
CLARK, WILLIAM M.
CURRY, WILLIAM A.
DICKERSON, C. J.
DRAIN, DAVID C. C.
ECHARD, WILLIAM
FISHER, JAMES S.
FORD, WILLIAM A.
GILBERT, JAMES M.
GILMORE, SAMUEL DAVIDSON
GLENN, JOSEPH
GORMAN, JOHN
GRIFFITH, WILLIAM H.
HALL, H. LEWIS
HALL, WILLIAM L.
HAMILTON, ALEXANDER C.
HARRIS, JOHN L.
HAYSLETT, BRADLEY
HAYSLETT, SERGEANT McDONALD
HICKS, WILLIAM H., JR.
HINKLE, WILLIAM H.
HOGUE, WILLIAM MADISON
HOSTETTER, LEWIS J. M.
HUGHES, DAVID E.
HUGHES, ROBERT H.
JONES, OLIVER
KIDD, JOHN PAXTON
KNICK, WILLIAM V.: 2nd Lt.
LAM, JOSEPH
LANG, ROBERT A.
LAWHORN, SAMUEL KIRKPATRICK
LONG, JAMES M.
MILLER, BENJAMIN FRANKLIN
MILLER, JOSEPH ANDREW
MOONEY, ABNER KIRKPATRICK
MUTERSPAUGH, DANIEL JAMES
MacMANAWAY, WILLIAM C.
McGUFFIN, JOHN WESLEY
OCHILTREE, DAVID LEECH
PARSONS, WILLIAM A.
PAXTON, MARTIN LUTHER
PAXTON, WILLIAM L.
ARMSTRONG, JOHN C.
AYERS, JOHN W.
AYRES, JAMES
BARGER, JOEL J.
BLACK, ANDREW D.
BLACK, JAMES S.
BLACK, WILLIAM L.
BRYAN, JOEL
BYERS, THOMAS C.
CALHOUN, EPHRIAM
CAMDEN, JAMES B.
CAMDEN, JOHN D.
CAMPBELL, MATHEW BRYAN: 2nd Lt.
CLARK, JAMES A.
CLARK, ROBERT
CROSSLEY, SYLVANIUS W.
DALE, SAMUEL K.
DIXON, WILLIAM A.
DRAIN, LEWIS C.
ENTSMINGER, JOHN AYERS
FISHER, JOHN ALEXANDER
GILBERT, ANDREW F.
GILMORE, ANDREW JACKSON
GLENN, JAMES MADISON
GLENN, ROBERT J.
GRANT, WILLIAM H.
HALL, E. A.
HALL, JOSEPH D.
HALL, WILLIAM T.
HARRIS, ALEXANDER
HAYSLETT, BENJAMIN FRANKLIN
HAYSLETT, RICHARD M.
HAYSLETT, WILLIAM, SR.
HILL, C. P.
HOGUE, SAMUEL E.
HOSTETTER, HENRY
HOSTETTER, MORGAN
HUGHES, LEWIS
JOHNSON, ROBERT M.
KIDD, CLIFFORD C.
KNICK, JOSEPH C.
LACKEY, WILLIAM HARVEY: Ensign
LAMBERT, ELBINE
LAWHORN, JOSEPH K.
LAYNE, JOSEPH
LONG, WILLIAM P.
MILLER, JOHN PRESSLY
MILLER, JOSEPH CLOYD
MOORE, BENJAMIN FRANKLIN
MUTERSPAUGH, JOHN
McDANIEL, ROBERT M.
NEWCOMER, WILLIAM S.
OCHILTREE, THOMAS ALEXANDER
PAXTON, JOSEPH SAMUEL: 1st Lt.
PAXTON, SAMUEL WASHINGTON: Captain
PHILLIPS, JOSEPH SOLOMON

PLYBON, JOHN
REESE, JOHN A.
REID, JAMES ALEXANDER
REID, WILLIAM HENRY (1)
REID, WILLIAM HENRY (2)
REID, WILLIAM H.
RILEY, PATRICK H.
RULEY, JOHN FRANKLIN
SCOTT, CHARLES A.
SCOTT, THOMAS F.
SELPH, BENJAMIN DANIEL
SELPH, WILLIAM JACKSON
SHAFER, ROBERT PRESTON GEORGE
SHAFER, SAMUEL JACKSON
SHAFFER, JEREMIAH HENRY
SHELTON, THOMAS A.
SHEPERDSON, ALFRED E.
SHORT, WILLIAM
SILVEY, JAMES M.
SIMPSON, CHARLES POWHATAN
SIMPSON, JOHN J.
SIMPSON, WILLIAM DAVID
SIRON, SIMON M.
SMITH, ANANIAS
SMITH, JAMES W. (1)
SMITH, JAMES W. (2)
SMITH, JOHN A.
SMITH, JOSEPH
SMITH, THOMAS
STOCKDALE, JOHN H.
THOMAS, FENDALL E.
TINSLEY, GEORGE R.
TUCKER, GARRETT L.
TYGRETT, WILLIAM ROBERT
VANHORN, SAMUEL
VESS, WILLIAM H.
WALLACE, JAMES WILLIAM M.
WATKINS, JOHN K.: 2nd Lt.
WELLS, CHRISTOPHER C.
WELLS, JAMES HENRY
WEST, WILLIAM C.
WHITE, JAMES M.
WHITTEN, JOHN W.
WILHELM, WILLIAM
WILKINSON, JOHN ALFRED
WILKINSON, WILLIAM WESLEY
WISEMAN, JACOB ADAM
WISEMAN, JOHN PETER
WOOD, JAMES A.
WOODY, HENRY
ZOLLMAN, ALEXANDER McCORKLE
ZOLLMAN, JOHN WILLIAM
ZOLLMAN, MADISON

COMPANY C

McDANIEL, CEOREIM JAMES
WOOD, THOMAS H.

COMPANY G

LYNN, ROBERT R.

COMPANY H

INGRAM, ALEXANDER
INGRAM, HUGH
INGRAM, JOSEPH W.
ZIMBRO, WILLIAM T.

COMPANY I

McKEE, JAMES MOFFETT

COMPANY K

SMITH, JOHN A.

KERR'S CREEK CONFEDERATES, COMPANY G, 58TH VIRGINIA INFANTRY

This company was organized in the community for which it was named during June-July 1861. The unit was sworn into Confederate service at Staunton on August 1, 1861. For the complete history of this company and the men who served in it, see the forthcoming volume, *58th Virginia Infantry.*

Colonel SAMUEL HOUSTON LETCHER
Lieutenant Colonel STAPLETON CRUTCHFIELD, JR.
Surgeon SAMUEL BROWN MORRISON
Captain & QM JAMES A. McCLUNG

AGNER, JOHN ALEXANDER, JR.
AILSTOCK, GEORGE
ALEXANDER, JOHN FRANKLIN
ARCHER, JOHN T.
ARCHER, PETER
BANE, SAMUEL R.
BENNINGTON, JOHN L.
BLACKBURN, D. B.
BLACKWELL, RICHARD
BROCKENBROUGH, ROBERT L.: Surgeon
CAMPBELL, CHARLES C.
CAMPBELL, JOHN A. J.
CARTER, JOHN A.
CONNER, FITZALLEN Y.
CRIST, RICHARD ABRAHAM
DAY, H. H.
DIXON, ROBERT J.
FARROW, PERRY: 2nd Lt.
FITZGERALD, SAMUEL
GAYLOR, JAMES MONTGOMERY
GRANT, ANDREW JACKSON
HALL, SAMUEL E.
HARRIS, DANIEL W.
HATTAN, MARK: Captain
HATCHER, JARVIS T.
HAYSLETT, ANDREW J. (1)
HAYSLETT, EZEKIEL (1)
HAYSLETT, JAMES M.
HENKLE, ADAM
HIGGINS, CHARLES
IRVINE, JOHN M.
KELLY, JOHN H.
LAM, JOHN S.
LAYMAN, W.
LINKSWILER, SAMUEL
MILLER, ANDREW P.
MILLER, SAMUEL HILEMAN
MOHLER, JAMES H.
MOHLER, WINSTON P.
MONTGOMERY, JOHN CRAWFORD
MORRIS, ELIHU BARCLAY
MORRIS, MARK
MORRISON, HENRY RUTHERFORD: 2nd Lt.
MUTERSPAUGH, GEORGE WASHINGTON
McDANIEL, JAMES HENRY
NUCKOLS, SILAS HENRY
PLOTT, JOHN A.
PLOTT, WILLIAM M.
REYNOLDS, MICHAEL C.
ROWSEY, MARION
SHAW, HENRY W.
SHAW, WESLEY W.
SMITH, BENJAMIN F.
SMITH, THOMAS, SR.
SPROUSE, JOSEPH L.
STEIN, D.
STOCK, A.
THOMAS, LEVI
THOMPSON, THOMAS JACKSON
TOLLEY, JAMES PRESTON
VESS, ANDREW JACKSON (2)
WANDLESS, STEPHEN H.
WILHELM, SAMUEL M.
WILSON, JAMES A.: 2nd Lt.
WISEMAN, JAMES F.
WYSAN, JAMES F.

AILSTOCK, ABSALOM
AILSTOCK, JOSEPH
ALEXANDER, W. M.
ARCHER, JOSEPH B.
ARMSTRONG, JAMES S.
BENNINGTON, JAMES NELSON
BENNINGTON, JOHN M.
BLACKBURN, RICHARD
BRADDS, JOHN ALEXANDER
BROWN, SAMUEL A.
CAMPBELL, CHARLES J.
CAMPBELL, ROBERT H.
CHAPLIN, JOHN: 1st Lt.
CRIST, HECTOR CROSON
DAVIDSON, CHARLES H.
DEACON, JOSEPH CRAWFORD
DONALD, ROBERT A.
FISHER, JOHN A.
FORD, JOSEPH
GRAHAM, ANDREW JACKSON
GRANT, SEATON L.
HAMILTON, ROBERT J.
HARTBARGER, FREDERICK STEPHEN
HAUBER, JOHN RUFUS
HAYSLETT, ANDREW
HAYSLETT, ANDREW J. (2)
HAYSLETT, EZEKIEL (2)
HAYSLETT, JAMES MADISON
HENKLE, PRESTON DUNLAP
HOSTETTER, GEORGE WILLIAM
JOLLY, CHARLES M.
LAM, JAMES W.
LAYMAN, HENRY
LINKSWILER, FRANKLIN
LOWMAN, SAMUEL
MILLER, HENRY M.
MILLER, WILLIAM
MOHLER, WILLIAM H.
MONTGOMERY, JAMES W.: 2nd Lt.
MOORE, JOHN PRESTON: Captain
MORRIS, JAMES D. W.
MORRIS, TILFORD B.
MORRISON, JAMES DAVIDSON: Captain
MUTERSPAUGH, WILLIAM CRAIG
McKEMY, ROBERT SAMUEL
PLOGGER, JAMES M.
PLOTT, JOSEPH N.
RAMSEY, NATHANIEL
RICHARDSON, JOHN C.: 2nd Lt.
RULEY, ROBERT TAYLOR
SHAW, SAMUEL B.
SHEPHERD, CHARLES
SMITH, JOHN H.
SMITH, THOMAS, JR.
STANLEY, LEWIS R.
STEIN, NIMROD
TEAFORD, GEORGE W.: 1st Lt.
THOMPSON, ALEXANDER
TOLLEY, CHARLES WILLIAM
VESS, ANDREW JACKSON (1)
VESS, HARVEY
WILHELM, JOHN H.
WILMORE, SEATON P.
WISEMAN, ANDREW J.
WISEMAN, JOHN A.

COMPANY A

DUDLEY, LINDSAY N.

COMPANY B

CHITTUM, GEORGE WASHINGTON
WATTS, JOHN HARVEY

CHITTUM, HENRY JEFFERSON

COMPANY C

EUBANK, JOHN W.
ROWSEY, JAMES
POOLEY, GEORGE W.
RICE, JOHN E.

COMPANY E

BAYS, HENRY H.
STATON, KILLIS E.
CUNNINGHAM, PEYTON R.

COMPANY F

BURKS, HUGHES N.: Captain
CHITTUM, STEPHEN E.
CLEMENTS, JOHN J.
LAIR, JOHN T.
PAINTER, WILLIAM
STATON, ANDREW L.
STATON, JAMES MADISON
CHITTUM, JOHN W.
CLEMENTS, JESSE
HARTLESS, PRESTON H.
NEESE, NEWTON
STATON, ABNER G.
STATON, ELIJAH S.
WHITESIDE, JAMES P.

COMPANY H

CAMPBELL, A. J.
BOOKER, GEORGE E.: Captain

COMPANY K

DOOLEY, JOHN L.

ROCKBRIDGE RANGERS

This company was organized at Lexington during May-June 1861. It served with General Henry A. Wise in western Virginia until October 1861, when it returned to Lexington and disbanded. Most of the men reenlisted in other units from Rockbridge County.

ANDERSON, WILLIAM I.
BELL, JOHN CYRUS
BRAFFORD, JAMES S.
BRANHAM, HORACE M.: Surgeon
CAMPBELL, WILLIAM C.
COPPER, THOMAS J.
DAVIDSON, ALBERT
DODD, ROBERT DUNBAR
FREEMAN, JAMES
HALL, JAMES C.
HARRIS, JAMES H.
HARVEY, JAMES ALEXANDER
IRVINE, JOHN
JOHNSON, ______
JUNKIN, WILLIAM FINLEY: Lt.
LACKEY, WILLIAM ALEXANDER
LEECH, WILLIAM BOLIVAR FINLEY
LUCAS, WILLIAM
MACKEY, JAMES
MANN, WILLIAM ALEXANDER
McMASTERS, SAMUEL McCONNELL
MOORE, RICHARD L.
NEWCOMB, ______
POAGUE, JAMES WILSON
SANDFORD, JAMES, III
SHAFER, SAMUEL JACOB
STEELE, WILLIAM
THOMAS, JAMES
TREVEY, DAVID A.
TURPIN, NASH
WILLIAMS, ______ Rev.
WILMORE, JAMES C.
WILSON, JAMES BROWN
WILSON, JOSEPH C.
WILSON, WILLIAM SAMUEL
ARMENTROUT, CORNELIUS
BOWYER, JOHN H.
BRAFFORD, ROBERT
CAMERON, JOHN: Lt.
CHRISTIAN, THOMAS W.
CRAIG, ROBERT S.
DAVIDSON, LEWIS C.: Captain
FIGGATT, JAMES
GRAHAM, EDWARD LACY
HAMILTON, ANDREW JACKSON
HARTSOOK, MARIUS Mc.
HAUGHAWOUT, JOHN W.
JENKINS, PHILIP
JORDAN, GEORGE W.: Captain
LACKEY, JAMES MORRISON
LEECH, HENRY MILLER
LEECH, WILLIAM CRAWFORD
LUCKESS, WILLIAM
MANN, GEORGE
McCORKLE, WILLIAM PHILANDER
MILLER, BENJAMIN F.
MOORE, WILLIAM T.
PATTERSON, JOHN M.
RHEA, SAMUEL T.
SHAFER, ARTHUR JACKSON
SITLINGTON, WILLIAM ALEXANDER
TIDBALL, THOMAS
TREVEY, CYRUS A.
TRIBBETT, WILLIAM W.
VESS, JOHN A.
WILLSON, WILLIAM S.
WILSON, ______
WILSON, JOHN
WILSON, ROBERT LUCIAN
YOUNG, JACOB

ROCKBRIDGE COUNTY SENIOR RESERVES

Companies A and B, 10th Battalion of Virginia Reserves, formerly the 4th Battalion Virginia Reserves, was organized on August 23, 1864. These two companies were reorganized from the three companies of Senior Reserves organized in Rockbridge County in April 1864 and served at Piedmont and during Hunter's raid to Lynchburg. These companies remained in active service until the end of the war. Partial muster rolls exist for some of these companies. Most of the names come from the author's files.

ADAMS, GEORGE
ADAMS, GEORGE W.
ADAMS, JOSEPH M.
AGNOR, JAMES ROBERT
AGNOR, JOHN H.
ALLISON, WILLIAM
ANDERSON, CHARLES H.
ARCHIBALD, JAMES
AUSTIN, WILLIAM RILEY
AYRES, JOHN REIPOGLE
AYRES, THOMAS G.
AYRES, WILLIAM LEWIS
BACON, ALGERNON SIDNEY: Captain
BAINE, GEORGE W.
BARGER, JOHN
BARGER, JOHN T.
BELL, JOHN MARSHALL
BENNINGTON, JOHN M.
BENSON, ARCHIBALD
BENSON, JOHN
BLACK, JOHN SAMUEL
BLAKEY, A. R.
BOON(E), ABRAHAM
BOON, LEWIS W.
BOWYER, JAMES HUBBARD
BRADLEY, JOHN J.
BROOKS, GEORGE W.
BROWN, JAMES M.
BROWN, JOHN ALEXANDER
BROWN, ROBERT H.
BUCHANAN, ROBERT EDWARD
BUMPASS, WILLIAM N.
BURGESS, MORGAN G.
BURKS, ALEXANDER H.
CAMPBELL, ROBERT
CHAMP, ANDREW JACKSON
CHAPIN, GEORGE
CHAPIN, WILLIAM
CHILDRESS, BENJAMIN
CLARKE, STROTHER W.
COFFMAN, GEORGE A.
COOLEY, JACOB F.
COX, JOHN S.
CRAFT, SOLOMON C.
CRIST, GEORGE W.
CRIST, JOHN J.
CUMMINGS, JOHN P.
CURRY, DAVID P.: Captain
DALE, MADISON
DAVIDSON, MADISON GILMORE
DAVIS, LEWIS PAYNE
DAY, JOHN S.
DEACON, DOUGLAS CRAWFORD
DECKER, SAMUEL HARVEY
DILL, JAMES M.
DIXON, SAMUEL
DODD, SHELTON
ECHOLS, EDWARD
EDMONDSON, JOHN M.
FIGGAT, RUTHERFORD H.
FIREBAUGH, BENJAMIN FRANKLIN
FIREBAUGH, DAVID
FORD, DAVID
FORD, JOSEPH
FOURMAN, J. H.
FRY, ZACHARIAH W.
FULLER, JACOB
FULWIDER, SAMUEL
GARING, JOHN F.
GAYLOR, THOMAS
GETTLE, H. H.
GIBSON, GEORGE W.
GLENN, PATTON G.
GOLD, SAMUEL
GOODBAR, ANDREW
GOODBAR, JAMES
GOODWIN, JOSIAH S.
GOOLSBY, ANDERSON
GORDON, ALEXANDER
GREEN, SAMUEL R.
GREENLEE, JOHN FRANKLIN
GROSE, JOHN
HALL, ANDREW H.
HALL, THOMAS L.
HAMILTON, DAVID H.
HAMILTON, JOHN GILBREATH: 2nd Lt.
HANGER, A. H.
HARDY, WILLIAM J.
HARRISON, RICHARD H.
HARRISON, WILLIAM HENRY
HARTMAN, HENRY THEOPHILUS
HAUGHAWOUT, JOHN W.
HAYSLETT, ANDREW
HECK, TILFORD B.
HENDERSON, JOHN F.
HEPLER, JOHN EDWARD
HERRING, WILLIAM J.
HESS, JOSEPH T.
HICKMAN, JACOB
HIGGINBOTHAM, W. T.
HILL, GEORGE W.
HILL, JONAS IRVINE
HITE, JAMES F.: Captain
HORN, JOHN
HOSTETTER, ANDREW
HOSTETTER, GEORGE WILLIAM
HOUSTON, N. J.
HUGHES, WILLIAM M.
HUMPHRIES, BENJAMIN FRANKLIN
HUNLEY, LEWIS V.
HUTCHESON, ROBERT STEELE, SR.: Captain
ICENHOWER, JOHNATHAN
JARVIS, ZACHARIAH
JOHNSON, MORTIMER HOWELL

JOHNSON, SAMUEL
JONES, WILLIAM E.
KELSO, EWING
KERR, THOMAS K.
KINNEAR, JOHN ALEXANDER
LACKEY, ANDREW H.
LAIRD, JOHN CALVIN
LAVELL, JOHN BEARD
LEWIS, B. F.
LOGAN, JOSEPH A.
LOGAN, ROBERT C.
LUCAS, JOHN A.
LUSK, ANDREW M.
MATCHETT, JAMES W.
MAYS, ALFRED T. G.
MILLER, ISAAC A.: Lt.
MILLER, JOHN
MILLER, SAMUEL STEELE
MOHLER, JAMES HENRY
MONTGOMERY, ROBERT: 2nd Lt.
MOORE, WILLIAM
MOORE, WILLIAM T.
MYERS, ALLEN
McCALEB, JOHN WESLEY
McCLURER, ROBERT CAMPBELL
McCORKLE, BENJAMIN F.
McCORKLE, WILLIAM H.
McCOWN, ALEXANDER JOHNSON, JR.
McCUTCHAN, WILLIAM MONTGOMERY
McMANAMA, THOMAS P.
OFFLIGHTER, HARRISON
PAXTON, JAMES J.
PECK, EDWIN
POWELL, J. SIMON
REES, NATHANIEL BROOKE
REYNOLDS, GEORGE
RISK, HARVEY
ROBERTSON, ROBERT
SAUNDERS, JAMES R.
SELBY, HENRY
SHANNON, JOHN BOYD
SHAW, JOHN ALEXANDER, SR.
SHEWEY, JOHN W.
SHIRLEY, ZECHARISH
SHUMATE, THOMAS
SLOUGH, WILLIAM T.
SMITH, JOHN
SMITH, WILLIAM H.
SORRELS, THOMAS JEFFERSON
STEELE, ALEXANDER TRIMBLE
STEELE, J. M.
SWINK, HENRY PHILANDER
TANKERSLY, SAMUEL ALEXANDER
TAYLOR, JAMES McDOWELL
TEMPLETON, JOHN MILLER: Captain
THOMPSON, JOHN: Captain & QM
TRIBBETT, ANDREW
TURPIN, H. C.: Captain
TYSINGER, GEORGE
WADE, ROBERT S.
WALLACE, CHARLTON
WALLACE, WILLIAM
WEBB, HENRY S.
WELSH, JAMES
WHITESELL, ELI
WHITESELL, JOHN H.
WILLSON, WILLIAM S.
WILSON, JAMES (1)
WILSON, SAMUEL
WOODS, JAMES
WREN, GEORGE W.
WRIGHT. JOHN D.

JONES, JAMES WILLIAM
JORDAN, WILLIAM HENRY
KERAN, HARRISON
KERR, ZACHARIAH A.: Captain
KIRKPATRICK, CHARLES BOYD: Captain
LACKEY, JAMES G.
LANCE, N. R.
LEECH, JAMES H.
LILLEY, WILLIAM
LOGAN, JOHN J.
LOWMAN, JOSEPH C.
LUDWICK, MOSES JACKSON
MARTIN, ISAAC D.
MATEER, WILLIAM COWAN
MILLER, ANDREW P.
MILLER, J. N.: 1st Lt.
MITCHELL, JAMES W.
MONTGOMERY, JAMES HALL
MOORE, ABNER WILSON
MOORE, WILLIAM M.: 2nd Lt.
MORRISS, WILLIAM
MORTER, DAVID
McCLUNG, JOSEPH ALEXANDER
McCLURER, WILLIAM PRESTON
McCORKLE, SAMUEL R.
McCORMICK, JOHN J.
McCUTCHAN, THOMAS K.
McMANAMA, ALEXANDER
McNUTT, JOHN RICE
PATTERSON, JOHN H.
PEARMAN, JOHN
POTTER, JOHN
RAMSEY, CALVIN
REID, WILLIAM H.
RHODES, WILLIAM A.
ROBERTSON, M. W.
ROBERTSON, THOMAS
SAVILLE, JOSEPH SKEEN
SHAFER, HENRY PHILIP
SHAW, DANIEL W.
SHAW, WILLIAM
SHIELDS, JOHN
SHORT, JOHN
SILERS, MATTHEW, Captain
SMITH, ALEXANDER
SMITH, OLIVER P.
SNIDER, ANDREW
SPROUSE, DAVID
STEELE, DAVID
SWEET, JACOB
SWISHER, HENRY
TARDY, WILLIAM JACKSON, Captain
TEAFORD, ELIJAH
TEMPLETON, WILLIAM W.
TOLLEY, JAMES PRESTON
TRUSLER, RICHARD
TYLER, JOHN J.
VEST, HIRAM H.
WALKER, FRANKLIN A.
WALLACE, SAMUEL PILSON: 1st Lt.
WALLS, THOMAS
WEBB, JAMES MILLER
WHEELER, SAMUEL C.
WHITESELL, GEORGE W.
WILKINSON, STEPHEN F.
WILSON, HENRY J.
WILSON, JAMES (2)
WILSON, WILLIAM
WOODS, WILLIAM W.
WRIGHT, ALEXANDER
WRIGHT, WILLIAM G.
ZOLLMAN, WILLIAM

HOME GUARDS OF ROCKBRIDGE COUNTY

These companies, for which no muster rolls exist, were made up of men too old, too young, or exempt for other reasons. The lists for the Lexington companies were found in the Lexington Town Minute Book. Other names came from the newspapers and the author's files.

ADAMS, CHARLES A.: Lexington Company

ALEXANDER, ARCHIBALD: Captain of Lexington Company

BARCLAY, JOHN WOODS: Captain of Lexington Company

BROWN, DANIEL: Lieutenant, Brownsburg Company

BURGESS, CYRUS A.: Lexington Company

CAMPBELL, ALEXANDER DOAK: Captain, Buffalo Cavalry

CAMPBELL, HENRY: Lexington Company

CAMPBELL, JOHN LYLE: Lexington Company

CAMPBELL, SAMUEL J.: Lexington Company

CHAMPE, ANDREW JACKSON: Lexington Company

CLOWES, AMOS K.: Lexington Company

CRAFT, SOLOMON C.: Lexington Company

CROKEN, JAMES H.: Lexington Company

DAVIDSON, GEORGE G.: Lexington Company

DAVIDSON, JOHN B.: Lexington Company

DEACON, JEFFERSON C.: Captain, Rapp's Mill Company

DOLD, WILLIAM: Lexington Company

DONALD, WILLIAM ALEXANDER: Captain, Fairfield Company

FORSYTHE, THOMAS: Lexington Company

FULLER, JACOB: Lexington Company

GARING, JOHN F.: Lexington Company

GIBSON, SAMUEL: 1st Lieutenant, South River Company

GILLOCK, SAMUEL: Captain, Lexington Company

GLASGOW, ALEXANDER McNUTT: Captain, South River Company

HAMILTON, JOHN W.: Lexington Company

HAMILTON, WILLIAM G.: 2nd Lieutenant, Fairfield Company

HARRIS, CARTER JOHN: Captain, Lexington Company

HARTSOOK, MARIUS Mc.: 2nd Lieutenant, Fancy Hill & Buffalo Forge Co.

HAUGHAWOUT, JOHN W.: Lexington Company

HAYNES, JOHN M.: 1st Lieutenant, Balcony Falls Company

HECK, TILFORD B.: Lexington Company

HENDERSON, HIRAM H.: Lexington Company

HILLIS, ROBERT J.: Lexington Company

HOLDEN, JOHN S.: 2nd Lieutenant, Oakdale Cavalry Company

HOPKINS, DAVID LAWRENCE: Lexington Company

JOHNSON, JAMES ALLEN: Rapp's Mill Company

JORDAN, GEORGE W.: Captain, Lexington Company

JUNKIN, WILLIAM FINNEY: Captain, Balcony Falls Company, Chaplain & Lieutenant Colonel, Rockbridge Home Guards Regiment

LEECH, JOHN ALEXANDER: 2nd Lieutenant, Collierstown Company

LEECH, WILLIAM BOLIVAR FRANKLIN: 1st Lieutenant, Oakdale Cavalry Company, 2nd Lieutenant, Natural Bridge Company

LINDSAY, J. W.: Lexington Company

MAITLAND, JAMES M.: Lexington Company

MARKS, RICHARD: Lexington Company

MIDDLETON, JOHN C.: Captain, Lexington Company

MITCHELL, ADAM WHITE: Balcony Falls Company

MYERS, JOHN H.: Lexington Company

McBRIDE, JAMES JACKSON: 1st Lieutenant, Brownsburg Company

McCAUL, JOHN R.: Lexington Company

McCOWN, ALEXANDER JOHNSON, JR.: Lexington Company

McCOWN, JOHN M. D.: Lexington Company

McCRUM, JAMES THADDEUS: 2nd Lieutenant, Lexington Company

McKEMY, JOHN H.: Captain, Kerr's Creek Company

NELSON, ALEXANDER LOCKART: Captain, Washington College Company

NORGROVE, HENRY: Lexington Company

PENDER, M. A.: Captain, Glenwood Cavalry Company

PLUNKETT, THOMAS B.: Lexington Company

POLE, JOHN G.: Captain, Lexington Company

RAMSEY, SIMON: Lexington Company

RHODES, FRANKLIN POINTS: Lexington Company

RHODES, WILLIAM A.: Captain, Lexington Company

RUFF, JACOB M.: 1st Lieutenant and Captain, Lexington Company

RUFFNER, WILLIAM HENRY: Lexington Company

SENSENEY, HARVEY K.: Lexington Company

SHIRLEY, ZECHARIAH: Lexington Company

STEELE, JOSEPH GRIGSBY: Captain, Lexington Company

STRAIN, DAVID ELDRED: Surgeon

STUART, JOHN H.: Major

SWITZER, SOLOMON: Lexington Company

STERRETT, JOHN DOUGLAS: Captain, Goshen Company

SHIELDS, GEORGE WASHINGTON: Lexington Company

TANQUARY, ALFRED B.: Lexington Company

TEAFORD, GEORGE C.: Lexington Company

THOMAS, JACOB: Lexington Company

TUCKER, JOHN W.: Lexington Company

VARNER, CHARLES VAN BUREN: Lexington Company

VARNER, NATHANIEL MACON: Lexington Company

WALLACE, WILLIAM: Lexington Company

WHITE, JAMES JONES: Captain, Lexington Company

WHITE, JOSEPH: Lexington Company

WHITE, ROBERT IRVINE: Captain, Lexington Company

WHITWELL, JOHN C.: Lexington Company

WILSON, HUGH LYLE: Lexington Company

WILSON, MATTHEW A.: South River Company

WRIGHT, JOHN G.: Lexington Company

WRIGHT, WILLIAM G.: Lexington Company

ROCKBRIDGE COUNTY MEN WHO SERVED IN UNITS RAISED IN OTHER COUNTIES AND OR STATES

ADAIR, JAMES LAFAYETTE: Marine, C.S.S. Tallahassee

ALEXANDER, OTHO: Co. L, 4th Va. Inf. and Captain, Co. K, 20th Va. Cav.

ALLEN, ALEXANDER M.: Co. D, 19th Va. Bn. Arty.

ALLEN, WILLIAM: Co. G, 19th Va. Inf.

ANDERSON, JOHN Y.: Co. C, 44th Va. Inf.

ANDERSON, MEREDITH: Co. D, 21st Va. Inf.

ANGUS, ZEBNON P.: Co. G, 18th Va. Cav.

ARGENBRIGHT, MORDICH W. D.: Carpenter's Va. Arty.

ARCHER, JOHN P.: Capt. McGuffin's Co., 26th Va. Cav.

ARMENTROUT, CHARLES E.: Co. D, 2nd Va. Inf.

ARNOLD, JACOB WYATT: Captain and QM, 40th Va. Bn. Cav.

AYERS, WILLIAM C.: Marquis' Boys Battery, Staunton.

AYRES, JOHN EDWARD: Co. A, 47th Bn. Va. Cav.

BABER, RICHARD STINBERG: Co. C, 39th Bn. Va. Cav.

BAILEY, JAMES M.: Co. E, 33rd Va. Inf.

BAILEY, WILLIAM H.: Huger's Va. Arty.

BAILIE, WILLIAM H.: Martin's Va. Arty.

BALDWIN, JOSEPH SALLING: McClanahan's Va. Horse Artillery

BANKER, VAN BUREN: Carpenter's Va. Arty.

BARE, WILLIAM A. B.: McNeill's Va. Rangers

BARGER, HENRY: Co. C, 34th Va. Inf.

BEARD, WILLIAM A.: 1st La. Inf.

BECKNER, JAMES O.: Botetourt Artillery

BECKNER, WILLIAM: Co. H, 28th Va. Inf.

BENNINGTON, ALEXANDER FRANKLIN: Petersburg Light Arty.

BLACK, HENRY ARMENTROUT: Co. I, 12th Va. Cav. and Co. K, 20th Va. Cav.

BLACK, JOHN McALLISTER: Co. C, 20th Bn. Va. Arty.

BLANTON, WILLIAM SOUTHALL: Co. D, 21st Va. Inf.

BOBBITT, MILTON LEFTWICH: Co. F, 3rd Ala. Inf.

BOLEY, WILLIAM HENRY: Bedford Arty.

BONDURANT, ALEXANDER JORDAN: Captain, Home Guards Co. and Reserves, Buckingham Co.

BOWEN, WILLIAM T.: Charlottesville Arty.

BOWYER, JAMES H.: Co. G, 2nd Va. Cav. and Surgeon.

BRADDS, HEZEKIAH: Co. C, 60th Ohio Inf. U.S.

BRADLEY, A. G.: 25th S.C. Inf.

BRAFFORD, JAMES E.: 10th Va. Cav.

BRAFFORD, ROBERT S.: Co. C, 34th Va. Inf.

BRAFFORD, STARKE: C.S. Navy

BRAGG, JOHN KIDD: Captain Massie's Va. Arty.

BRISINDINE, CLEMENT: Botetourt Arty.

BRITTON, JOHN NEWTON: Captain Brown's Co., 39th Bn. Va. Cav.

BROCHUS, JAMES: Co. I, 12th Va. Cav.

BROGAN, J. EDWARD: Co. K, 60th Va. Inf.

BROOKS, JAMES M.: Co. E, 3rd C.S. Engineers

BROUGHMAN, FRANCES: Co. F, 60th Va. Inf.

BROWN, JAMES NELSON: Marine, C.S.S. Albemarle

BROWN, PERRY J.: Marquis' Boys Battery, Staunton.

BROWN, ROBERT WASHINGTON: Co. A, 1st Va. Bn. Inf. (Irish)

BROWNLEE, EDWARD GOULD: McClanahan's Va. Horse Arty.

BRUCE, JAMES W.: Richmond Fayette Arty.

BRUCE, ROBERT A.: Co. C, 34th Va. Inf.

BRYANT, JOHN H.: Co. I, 20th Va. Cav.

BUFFINGTON, E. F.: Co. D, 10th Va. Cav.

BURGER, JOEL: Co. D, 60th Va. Inf.

BURKE, JOHN: Co. C, 44th Va. Inf.

BURKS, RICHARD HORSLEY: Lt. Col., 12th Va. Cav.

BURKS, THOMAS GOOLSBY, JR.: Co. I, 12th Va. Cav.

BYERS, JAMES DAVIS: Co. I, 8th Va. Cav.

CALE, JAMES W.: Marquis' Boys Battery, Staunton

CALVERT, ALEXANDER H.: McClanahan's Va. Horse Arty.

CAMDEN, WILLIAM: Co. I, 17th Va. Cav.

CAMERON, GEORGE HUGH: Co. H, 18th Va. Cav.

CAMPBELL, ARCHIBALD: Nelson Arty.

CAMPBELL, BARTHOLOMEW: Reeves Va. Btty.

CAMPBELL, JAMES DORMAN: Co. E, 1st Texas Inf.

CAMPBELL, JOHN THOMAS: McClanahan's Va. Horse Arty.

CAMPBELL, MATTHEW J.: Co. I, 8th Va. Inf.

CAMPBELL, ROBERT A.: 10th Va. Cav.

CAMPBELL, WILLIAM C.: Co. F, 22nd Va. Inf.

CAMPER, WILLIAM HENRY: Co. A, 28th Va. Inf.

CANODAY, THOMAS: Co. C, 35th Va. Bn. Cav.

CARTER, BERRYHILL McLEAN: 43rd Va. Bn. Cav.

CARTER, FREDERICK K.: Co. A, 22nd Va. Inf.

CASH, BENJAMIN M.: Nelson Arty.

CASH, JAMES CLARK: Co. C, 10th Va. Inf.

CATLETT, JOHN F.: Co. H, 28th Va. Inf.

CHAPMAN, JOHN: Captain Robertson's Co., 1st Va. Bn. Inf. (Irish)

CHILDRESS, DAVID DILLARA: Co. C, 39th Bn. Va. Cav.

CHILES, WILLIAM H.: Fitz Lee's Cav., from Louisa Co.

CHITTUM, JAMES A.: Co. C, 6th Va. Cav.

CHITTUM, NATHANIEL ANDERSON: Co. D, 28th Va. Inf.

CHRISMAN, ISAAC: Co. F, 54th Va. Inf.

CHRISTIAN, THOMAS W.: Co. D, 23rd Va. Bn. Inf.

CLARK, LEWIS F.: Co. F, 5th Va. Cav.

CLEMENTS, JOHN J.: 49th Va. Inf.

CLEMMER, GEORGE LEWIS: Co. I, 62nd Va. Inf.

CLINDINST, JOHN W.: 10th Va. Inf.

COFFEY, ALFRED: Latham's Va. Arty.

COFFEY, JORDAN: Co. I, 49th Va. Inf.

COFFEY, REUBEN W.: Reeves Va. Arty.

COLEMAN, SAMUEL I.: Co. I, Col. Booker's Regt. Va. Reserves.

COLONNA, BENJAMIN AZARIAH: New Market Cadet, Marquis' Boys Battery, Staunton, Chew's Horse Arty, Captain, 1st Foreign Bn., 1st Conf. Regt.

COLSTON, RALEIGH EDWARD: Colonel, 16th Va. Inf., Brigadier General

CONNER, DENNIS: 25th Bn. Va. Res.

CONNER, GEORGE: Fauquier Arty.

CONNER, WILLIAM ROBERT: Captain Avis' Provost Guard, Staunton.

CONNEVEY, SAMUEL MORANDA: 2nd Lt., Co. K, 24th Va. Inf.

COPPER, JOHN DAVID: Co. E, 20th Va. Inf.

COUCH, CHARLES H.: Co. G, 2nd Va. Cav.

COX, BAYLOR: Amherst Arty.

COYNER, JAMES A.: Co. G, 18th Va. Cav.

CRAWFORD, GEORGE WASHINGTON: Bedford Light Arty.

CRAWFORD, JAMES LEITCH: Chew's Horse Arty.

CRESS, JOHN GRAHAM: Captain, Cobb's Legion. Major, 8th Ga. Cav.

CROSS, JOHN A.: Co. F, 50th Va. Inf.

CULIN, GEORGE W.: Co. A, 19th Va. Inf. and Co. G, 25th Bn. Va. Res.

CUMMINGS, HARVEY: Carpenter's Va. Arty.

CUMMINS, HENRY RUFFNER: Lt. Ark. Inf.

CUNNINGHAM, JOHN R.: Co. G, 51st Va. Inf.

DABNEY, WILLIAM A.: 63rd Ga. Inf.

DADEN, JAMES: Eubank's Va. Arty.

DAILEY, JAMES: Whittington's Va. Arty.

DALE, JOHN W.: Co. G, 42nd Va. Inf.

DAVIDSON, ALBERT L.: Letcher Arty. and 1st Lt. & AAG Gen. Robert Preston.

DAVIDSON, CHARLES ANDREW: Captain, Co. E, 1st Va. Bn. Inf. (Irish)

DAVIDSON, CHARLES H.: Co. K, 22nd Va. Inf.

DAVIDSON, HENRY GAMBLE: Surgeon, 3rd and 5th Va. Inf.

DAVIDSON, JOHN B.: Co. C, 10th Va. Inf.

DAVIDSON, JOHN GREENLEE: Captain, Letcher Arty.

DAVIDSON, WILLIAM WEAVER: Letcher Arty. and 1st Lt., Co. I, 26th Bn. Va. Inf.

DEACON, ANDREW JAMES: Floyd's State Line Troops

DEACON, WILLIAM DOUGLAS: Co. E, 3rd C.S. Engineers

DEMPSEY, JOHN J.: Carpenter's Va. Arty.

DILLON, EDWARD: Colonel, 2nd Miss. Cav.

DIXON, THOMAS J.: Co. H, 2nd Va. Cav.

DONALD, ALEXANDER JOSEPH: Co. C, 20th Va. Inf.

DOOLEY, JOHN M.: Co. I, 22nd Va. Inf.

DORMAN, JAMES BALDWIN: Major, 9th Va. Inf.

DORMAN, WILLIAM BOLIVAR: Captain, 50th Va. Inf.

DOUGLASS, BURTON. R.: Co. G, 18th Va. Cav.

DRAWBOND, WILLIAM HENRY: Co. F, 60th Va. Inf.

DUDLEY, GEORGE W.: Co. K, 8th Va. Cav. and Jackson's Va. Horse Arty.

DUDLEY, JAMES C.: Co. F, 12th Va. Cav.

DUERSON, S. K.: Crenshaw's Va. Arty.

DULANEY, JOHN T.: Co. E, 5th Tex. Inf.

DUNAWAY, WILLIAM: Co. B, 19th Va. Inf.

DUNCAN, VALENTINE: Co. E, 5th Va. Cav.

DUNLOP, JOHN THOMAS: Co. G, 7th Va. Cav.

EARMAN, JOSEPH ALPHEUS: Co. I, 39th Bn. Va. Cav.

ECHARD, JOHN C.: Co. C, 20th Va. Inf.

EDMONDS, GEORGE DOROMAN: Co. H, 4th Va. Cav.

EDWARDS, OSCAR: 34th Va. Inf.

EDWARDS, THOMAS W.: 60th Va. Inf.

EDWARDS, WILLIAM H.: Co. A, 35th Bn. Va. Cav.

ENGLEMAN, JAMES WILLIAM: Co. E, 46th Va. Bn. Cav.

ENTSMINGER, DAVID E.: (1) Co. E, 46th Bn. Va. Cav.

ENTSMINGER, DAVID E.: (2) Letcher Arty.

ENTSMINGER, SAMUEL: 37th Bn. Va. Cav.

ESTILL, JOHN MILLER: Surgeon, 51st Va. Inf.

ESTILL, WILLIAM G.: 36th and 34th Bn. Va. Cav.

FALLS, JAMES C.: Botetourt Arty.

FALLS, JOSEPH: Captain, Whitehead's Co., 2nd Va. Cav.

FARMER, R. S.: Co. G, 34th Va. Inf.

FIGGATT, FERDINAND W.: Carpenter's Va. Arty.

FIGGATT, JOHN CLAY: Eubank's Va. Arty.

FIREBAUGH, ROBERT DUNLAP: Co. I, 82nd Va. Inf.

FIREBAUGH, SAMUEL A.: Co. H, 10th Va. Inf.

FISHER, DAVID F.: Co. E, 1st Va. Bn. Inf. (Irish)

FISHER, FRANKLIN A.: Co. B, 62nd Va. Inf.

FITZGERALD, A. F.: Co. D, 36th Va. Bn. Cav.

FITZGERALD, ALEXANDER J.: Co. F, 28th Va. Inf.

FITZGERALD, ALTA H.: Rieves Va. Arty.

FITZGERALD, JOHN: Co. D, 36th Va. Bn. Cav.

FITZGERALD, W. C.: Co. D, 36th Va. Bn. Cav.

FITZGERALD, WILLIAM: Co. D, 36th Va. Bn. Cav.

FITZPATRICK, ALEXANDER BRECKENRIDGE: Nelson Arty. and Co. F, 28th Va. Inf.

FLEMING, JOHN: Allen's Va. Arty.

FLETCHER, LUCIAN: Lee Va. Arty.

FLOYD, SAMUEL: Co. I, 49th Va. Inf.

FORD, WILLIAM: Co. C, 8th Va. Inf.

FORTUNE, JOHN: Botetourt Arty.

FOSTER, JOHN W.: Co. C, 18th Va. Inf.

FOSTER, THOMAS F.: Co. I, 7th Va. Cav.

FOX, JAMES R.: Co. D, 42nd Va. Inf.

FOX, WILLIAM BOWLES: Co. G, 46th Va. Inf.

FRANKLAND, VALENTINE T.: Co. D, 60th Va. Inf.

FULTZ, JOSEPH: Chew's Va. Horse Arty.

FULWIDER, JAMES W.: Co. K, 20th Va. Bn. Cav.

GABBERT, DAVID BITTLE: Co. C, 39th Va. Bn. Cav.

GALLOWAY, LEAKE: Powhatan Va. Arty.

GASSMAN, JACOB: Co. F, 7th Va. Cav.

GAYLOR, THOMAS M.: Co. D, 26th Va. Cav.

GILBERT, ROBERT NORVAL: Co. H, 19th Va. Inf.

GILES, J. J.: Lampkins Va. Bty.

GILES, SAMUEL N.: Co. G, 7th Va. Cav.

GILHAM, WILLIAM: Colonel, 21st Va. Inf.

GILHAM, JULIUS FRANCIS: C.S. Engineers

GILLIAM, RICHARD BENJAMIN: 57th Va. Inf.

GILLOCK, JOHN J.: Co. F, 12th Va. Cav.

GLEASON, JOHN J.: 43rd Va. Bn. Cav.

GOFF, WILLIAM ALEXANDER: Co. C, 42nd Va. Inf.

GOLD, ROBERT P.: Co. B, 45th Va. Inf.

GORMAN, JAMES P.: Richardson's Va. Arty.

GORRELL, BENJAMIN HARVEY: Lt., Sturdivant's Va. Arty., Co. B, 13th Va. Inf.

GRAHAM, JOHN ALEXANDER: Surgeon, 42nd Va. Inf.

GRANT, JAMES: 1st Va. Bn. Inf. (Irish)

GRANT, PAUL N.: Co. C, 34th Va. Inf.

GRANT, WILLIAM H.: Co. C, 34th Va. Inf.

GREEN, JAMES LANE: C.S.N.

GREEN, JOHNATHAN: 21st Va. Inf.

GREENLEE, ELISHA GRIGSBY: Surgeon, 2nd Ky. Inf.

GRIFFIN, JACK S.: Co. B, 3rd Va. Inf.

GRIGSBY, JOHN: Colonel, 6th Ky., Brig. Gen. and I.G. under Joseph E. Johnston

GRINSTEAD, JACOB VALENTINE: Smith's Bn., Morgan's Ky. Cav.

GROVE, JOSHUA M.: Co. C, 12th Va. Cav.
GUGGENHIMER, MORRIS: Co. C, 2nd Va. Cav.
HACKWORTH, LEWIS C.: Latham's Va. Arty.
HALL, JOHN TRENT: Co. H, 46th Va. Inf.
HAM, JOSEPH F.: Marquis' Boys Battery, Staunton
HAMILTON, AUGUSTUS HOUSTON: Chapman's Va. Arty.
HAMILTON, GEORGE WASHINGTON: Co. I, 49th Va. Inf.
HAMILTON, JOHN A.: Co. G, 6th Va. Bn. Inf.
HAMILTON, JOHN LUCIAN: Co. G, 19th Va. Cav.
HAMILTON, ROBERT CLAY: Co. E, 46th Va. Bn. Cav.
HAMILTON, ROBERT L.: Captain Charles Mundy's Co. of Arty.
HANGER, ANDREW SANFORD: Co. C, 17th Va. Cav.
HARDIN, MARK BERNARD: Major, 16th Va. Bn. Arty.
HARLOW, ANDREW MORTIMER: Albemarle Arty.
HARLOW, JAMES M.: Co. H, 57th Va. Inf.
HARLOW, JOHN WILLIS: Co. I, 46th Va. Inf.
HARLOW, WILLIAM ALBERT: Co. F, 44th Va. Inf.
HARMAN, J. W.: Co. E, 24th Va. Inf.
HARRAH, JOHN A.: Co. D, 22nd Va. Inf.
HARRIS, JAMES ROBERT JACKSON: Co. G, 54th Va. Inf.
HARRIS, JOHN RICE J.: Co. F, 23rd Va. Cav.
HARRIS, JOSEPH C.: Co. B, 23rd Va. Cav.
HARRIS, SAMUEL D.: Kirkpatrick's Va. Arty.
HARRIS, WILLIAM H.: Lt., Co. C, 34th Va. Inf.
HARTLESS, BENJAMIN: Co. I, 19th Va. Inf.
HARVEY, JOHN H.: Co. C, 34th Va. Inf.
HARVEY, ROBERT BRECKINRIDGE: Captain, 6th Tex. Inf.
HATCHER, WILLIAM LOGWOOD: Co. B, 12th Va. Cav.
HATCHER, WILLIAM RICHARD: Co. G, 34th Va. Inf.
HAWKINS, JAMES WOODS: Co. F, 28th Va. Inf.
HAYSLETT, JAMES MADISON: Co. A, 47th Va. Bn. Cav.
HAYNES, JOHN H.: Co. C, 34th Va. Inf.
HAZLEWOOD, JOHN A.: Co. C, 2nd Va. Cav.
HEIZER, JAMES FRANCIS: McClanahan's Va. Horse Arty.
HENDERSON, JOHN LETCHER: Captain Avis' Co., Provost Guard, Staunton.
HENDERSON, OCTAVIUS CAZENORE: Captain, Co. D, 1st Va. Bn. Inf. (Irish)
HENKLE, ADAM: Co. I, 20th Va. Cav.
HENKLE, ANDREW P.: Co. I, 20th Va. Cav.
HENSLEY, SAMUEL E.: Co. E, 34th Va. Inf.
HEFLER, SAMUEL M.: Co. G, 18th Va. Cav.
HICKMAN, ALEXANDER: Co. C, 34th Va. Inf.
HICKMAN, ALFRED: Co. C, 34th Va. Inf.
HICKMAN, GEORGE: Johnston's Va. Arty.
HICKMAN, GEORGE T.: Co. C, 34th Va. Inf.
HICKMAN, LEWIS: Captain, Co. D, 20th Va. Cav.
HICKMAN, MATTHEW A.: Co. C, 34th Va. Inf.
HIGGINBOTHAM, THOMAS JOSEPH: Co. E, 2nd Va. Cav.
HILL, JACOB ISAAC: Captain & QM, 31st Va. Inf. & Early's Brigade
HINTY, WILLIAM HENRY: 2nd Lt., Co. E, 19th Va. Cav.
HITE, M. M.: Co. A, 39th Bn. Va. Cav.
HODGES, THOMAS L.: Co. H, 17th Va. Cav.
HODGES, VERNON E.: Co. G, 1st Va. Inf.
HOGAN, TIMOTHY: St. Bride's Arty. & Co. I, 38th Va. Inf.
HOGSETT, J. McD.: Co. I, 1st Va. Inf.
HOGSETT, HENRY L. V.: Co. C, 20th Va. Cav.
HOGSHEAD, PRESTON BAILEY: 2nd Va. Inf.
HOGSHEAD, SAMUEL McC.: 19th Va. Cav.
HOLT, GEORGE W.: Co. C, 34th Va. Inf.
HOOK, EDWIN E.: Co. D, 60th Va. Inf.
HOPKINS, HENRY ST. GEORGE: Surgeon, 2nd Va. Bn. Inf. and 3rd Va. Bn. Arty., Staff of Col. Nelson of Arty.
HOUSTON, ARCHIBALD WOODS: Lt., Charlottesville Arty.
HOUSTON, HORACE MYERS: Lt., 4th Tenn. Inf.
HOUSTON, WILLIAM PAXTON: Lt., Lowry's Va. Arty.
HOUSTON, WILLIAM WILSON: Chaplain, McIntosh's Bn. Arty.
HOWARD, JOHN C.: Co. C, 34th Va. Inf.
HUFFMAN, DAVID S.: McClanahan's Va. Horse Arty.
HUFFMAN, GEORGE D.: Co. D, 23rd Va. Bn. Inf.
HUFFMAN, JESSIE H.: Co. D, 23rd Va. Bn. Inf.
HULL, JOHN McKEE: Co. K, 22nd Va. Inf.
HULL, JOSHUA: 22nd Va. Inf.
HUMPHREYS, HOWARD A.: 3rd Ark. Cav.
HUMPHRIES, SAMUEL: Marquis' Boys Battery, Staunton
HUMPHRIES, WILLIAM NADEL: Reeves' Va. Arty.
HUNTER, ROBERT: Co. C, 10th Va. Inf.
HUNTER, THOMAS ALLEN: 1st Lt., Botetourt Arty.
HUTCHENS, SAMUEL W.: Co. G, 10th Va. Inf.
ICENHOWER, JAMES S. C.: Co. D, 60th Va. Inf.
ILLIG, JOHN ST. STEPHEN: Co. I, 1st Va. Inf.
IMBODEN, JOHN ALEXANDER RUFF M.: Staunton Arty., McClanahan's Va. Horse Arty., 18th Va. Cav., 62nd Va. Inf., 23rd Va. Cav., Captain and AAG, Imboden's Brigade
ISAACS, WILLIAM H.: Co. C, 34th Va. Inf.
JACKSON, ALFRED HENRY: Lt. Colonel, 31st Va. Inf.
JACKSON, JOHN ANDREW: Crenshaw's Va. Arty.
JACKSON, THOMAS JOHNATHAN: Lt. General "Stonewall"
JEFFRIES, DAVID H.: Co. C, 17th Va. Cav.
JOHNSON, GEORGE D.: Co. C, 34th Va. Inf.
JOHNSON, JOHN J.: Co. C, 34th Va. Inf.
JONES, EDWARD J.: Amherst Arty.
JONES, TAZEWELL L.: Co. A, 60th Va. Inf.
JONESSHIRE, P.: Co. B, 19th Va. Inf.
JORDAN, JOHN WILLIAM: Jackson's Va. Horse Arty.
KAYTON, THOMAS J.: Co. K, 26th Va. Cav.
KERICOFF, L.: Co. E, 46th Va. Cav.
KERR, RICHARD ADDISON: Co. A, Marine Bn.
KINNEAR, JOHN ALEXANDER: Co. B, 2nd Va. Cav.
KIRBY, WILLIAM BOWERS: Albemarle Arty. & Co. C, 34th Va. Inf.
KIRKPATRICK, GEORGE RENNICK: Carpenter's Va. Arty.
LACKEY, JAMES SHAFER: Co. D, 19th Va. Cav.
LACKEY, JOHN H.: (1) Co. F, 12th Va. Cav.
LACKEY, JOHN H.: (2) Co. E, 46th Bn. Va. Cav.
LAKE, JOHN HARVEY: Co. I, 12th Va. Cav.
LAM, ISAAC DONALD: Co. I, 10th Va. Inf. & Co. C, 22nd Va. Inf.
LAWHORN, WILLIS: Nelson Arty.
LECKEY, THEOPHILIUS SAUNDERS: 22nd Tenn. Inf.
LEE, EDWIN GRAY: Colonel, 33rd Va. Inf. and B. Gen.
LEE, GEORGE WASHINGTON CUSTIS: Major General
LEE, ROBERT EDWARD: *General*
LEECH, JAMES RODGERS: Co. K, 20th Va. Cav.
LEECH, THOMAS LACKEY: Co. K, 22nd Va. Inf.
LEECH, WILLIAM PHILANDER: Co. K, 22nd Va. Inf.

LEGARD, CHARLES: Co. C, 19th Va. Cav.

LEIGHTON, LEWIS EDWARD: Co. C, 34th Va. Inf.

LEIGHTON, ROBERT L.: Co. C, 34th Va. Inf.

LEIGHTON, WILLIAM R.: Co. C, 34th Va. Inf.

LENTZ, JOHN HENRY: 1st Va. Bn. Inf. (Irish)

LEYBURN, GEORGE LACON: Lt., 34th Va. Inf.

LILLY, DAVID F.: Co. F, 27th Va. Bn. Cav.

LINDSAY, PAUL McNEEL: Co. G, 31st Va. Inf. & Co. G, 18th Va. Cav.

LINDSAY, SAMUEL CARL: Co. G, 31st Va. Inf.

LINK, DAVID: Rice's Va. Arty.

LINKENHOKER, JOHN G.: Co. C, 34th Va. Inf.

LIPSCOMB, CLEMENT JORDAN: Co. C, 18th Va. Inf.

LITTLE, SAMUEL M. C.: Dabney's Va. Arty.

LOCKRIDGE, EDWARD FRANKLIN: 20th Va. Cav.

LOGAN, CHARLES H.: Co. K, 23rd Va. Cav.

LOGAN, JOHN W.: Co. F, 31st Va. Inf.

LOKER, WILLIAM H.: Co. F, 12th Va. Inf.

LOTTS, JAMES FRANKLIN: Carpenter's Va. Arty.

LOVE, HAMILTON JONES: Co. G, 6th N.C. Inf.

LUCADO, ABNER: Co. B, 4th Va. Inf.

LUCAS, WILLIAM HENRY: Co. E, 1st Va. Bn. Inf. (Irish)

LUCK, MARCELLES FLUMAN: Crenshaw's Va. Arty.

LUNSFORD, JAMES F.: Staunton Arty.

LYLE, GEORGE TATE: Captain, Co. K, 25th Va. Cav.

MACKEY, JOHN McBRIDE: Carpenter's Va. Arty.

MADDOX, NELSON W.: Reeves Va. Arty.

MADDOX, NOTLEY W.: Reeves Va. Arty.

MADISON, JOHN W.: Co. I, 5th Va. Cav.

MAHONEY, EDWARD M.: 28th Va. Inf.

MANKEY, J. A.: Co. I, 22nd Va. Inf.

MANN, WILLIAM HODGES: Co. E, 12th Va. Inf.

MANOR, W. A.: 17th Bn. Va. Cav.

MANSPILE, ALLEN J.: Botetourt Arty.

MARKEY, FRANK ANTON: Staunton Arty.

MARTIN, ASBURY R.: Salem Arty.

MARTIN, REUBEN LAWRENCE: Carpenter's Va. Arty.

MARTIN, THOMAS R.: Co. F, 10th Va. Cav.

MASON, GEORGE WASHINGTON: Amherst Arty.

MASSIE, JAMES WOODS: Lt. Colonel, 51st Va. Inf., V.M.I. staff 62-65, Captain of Patrol, Lexington 1862, Colonel of Rockbridge Home Guards 1863.

MATHENY, GEORGE C.: Botetourt Arty.

MAYS, PAUL ALEXANDER: Latham's Va. Arty.

MAZINGO, C. T.: Co. G, 12th Va. Cav.

MERRIWETHER, GEORGE DOUGLAS: Co. G, 2nd Va. Cav.

MERRIWETHER, WILLIAM M.: 21st Va. Inf.

MILLER, LEVI: (Colored) Co. C, 5th Tex. Inf.

MILLER, DANIEL K.: Co. A, Va. Heavy Arty.

MILLER, JOHN F.: Kirkpatrick's Va. Arty.

MILLER, WILLIAM B.: Kirkpatrick's Va. Arty.

MINES, COLUMBUS W.: Co. E, 1st Va. Bn. Inf. (Irish)

MITCHELL, WILLIAM G.: Letcher Arty.

MONROE, ALEXANDER MARSHALL: Captain, Co. F, 2nd Va. Inf.

MONTEGUE, JAMES ROBERT: Carpenter's Va. Arty.

MONTGOMERY, ANDREW SCOTT: Carpenter's Va. Arty.

MONTGOMERY, E. D.: Co. F, 60th Va. Inf.

MONTGOMERY, SAMUEL: Co. C, 19th Va. Cav.

MONTGOMERY, SAMUEL: Co. K, 22nd Va. Inf.

MOORE, ABNER WILSON: Letcher Arty.

MOORE, AMBROSE B.: Co. H, 57th Va. Inf.

MOORE, ANDREW: Lt., Kanawha Rifles, 22nd Va. Inf. and Imboden's Cav., ADC Gen. D. H. Hill.

MOORE, ARMISTEAD B.: Co. C, 17th Va. Cav.

MOORE, DAVID HUMPHRIES: Co. C, 42nd Va. Inf.

MOORE, SAMUEL H.: Co. B, 7th Va. Cav. and McNeill's Rangers

MORAN, ISAAC K.: Co. D, 14th Va. Inf.

MORGAN, WILLIAM HENRY: Captain, 21st Va. Inf.

MORRIS, DAVID MILLER: Co. I, 20th Va. Cav.

MORRIS, WILLIAM C.: Co. C, 34th Va. Inf.

MORRIS, WILLIAM TAYLOR: Co. I, 20th Va. Cav.

MORRISON, HENRY RUFFNER: Adjutant, 23rd Va. Bn. Inf.

MORRISON, JAMES LUTHER: Allen's Va. Arty.

MORRISON, SAMUEL BROWN: Co. H, 17th Va. Cav., Surgeon, McCausland's Brigade, Chief Surgeon, Ewell's Division

MORRISON, W. T.: 20th Va. Cav.

MOXLEY, GEORGE W.: Co. C, 34th Va. Inf.

MURPHY, MICHAEL: Co. D, 19th Va. Inf.

MURRAY, J. OGDEN: Co. K, 7th Va. Cav.

MUSE, JAMES BLACKERBY: 4th Ala. Inf.

McBRYDE, ROBERT J.: Co. G, 1st S.C. Cav.

McCANN, WILLIAM: Co. E, 1st Va. Bn. Inf. (Irish)

McCLURE, JOHN PILSON: Co. B, 23rd Va. Cav.

McCLURER, JOHN GRIGSBY: Co. B, 12th Va. Cav.

McCLURER, JOHN WILLIAM: Letcher Arty.

McCORKLE, JOHN S.: Amherst Arty.

McCORMACK, WILLIAM TATE: Captain Avis' Co., Provost Guard, Staunton

McCOY, JAMES WALKUP: Co. F, 33rd Va. Inf.

McCOY, MONTICELLO H.: Co. B, 4th Va. Cav.

McCULLOUCH, ROBERT L.: Captain, Danville Grays, 18th Va. Inf.

McCULLOUGH, JAMES: Co. A, 28th Va. Inf.

McCUNE, ROBERT H.: Co. D, 18th Va. Inf.

McDANIEL, WILLIAM HENRY HARRISON: Co. B, 31st Va. Inf.

McDONOUGH, WILLIAM W.: Co. A, 35th Bn. Va. Cav.

McFADDEN, ISAAC: 1st Ark. Cav. & 9th Ark. Inf.

McFADDEN, JOHN: Surgeon, 1st Ark. Cav.

McFARLAND, ROBERT: Captain, 4th Ala. Inf., Captain, Morgan's Ky. Cav. and on staff Gen. Patrick Cleburne

McGUFFIN, JOHN T.: Marquis' Boys Battery, Staunton

McGUIDY, W. T.: Co. I, 10th Va. Inf.

McLAUGHLIN, WILLIAM: Co. H, 19th Va. Cav.

McMANAMA, THOMAS P.: Captain Hart's Co., Engineers

McNEEL, JOHN ADAM: Co. F, 19th Va. Cav.

NEWCOM, WILLIAM H. H.: Botetourt Arty.

NICELY, ALEXANDER A.: Co. I, 22nd Va. Inf.

NICELY, ANDREW JACKSON: Co. F, 60th Va. Inf.

NICELY, EMANUEL: Co. G, 22nd Va. Inf.

NICELY, GEORGE A.: Co. G, 22nd Va. Inf.

NICELY, JAMES A.: Co. K, 20th Va. Cav.

NICELY, PETER: Co. G, 22nd Va. Inf.

NOEL, WILLIAM L.: Co. A, 2nd Va. Cav.

NORGROVE, WILLIAM H.: 2nd Lt., Botetourt Arty.

NORTON, WILLIAM BUNTON: Co. H, 29th Va. Inf.

NUCKOLS, JOSEPH J.: Eubanks Arty.

NUTTY, JOHN W.: Co. F, 60th Va. Inf.

OBENSCHAIN, JAMES PATTERSON: Co. A, 28th Va. Inf.

OFFLIGHTER, JOSEPH D.: Co. E, 36th Va. Inf.

OGDEN, BENJAMIN ABROSE: Co. I, 49th Va. Inf.

OSBORN, WILLIAM C.: Co. A, 6th Va. Inf.

OVERSTREET, E. V.: Co. H, 34th Va. Inf.

OVERTON, JAMES E.: Co. A, 20th Va. Bn. Arty.

OVERTON, JOHN H.: Co. A, 20th Va. Bn. Arty.

PAGE, GABRIEL FRANCIS: Co. I, 19th Va. Inf.

PAGE, WILLIAM NELSON, JR.: Co. E, 18th Va. Inf.

PAINTER, JOSEPH: Lampkin's Va. Arty.

PAINTER, WILLIAM PAXTON: Co. B, 4th Va. Cav.

PARKER, JAMES RUFF: Co. H, 17th Va. Cav.

PARKER, THOMAS BURNS: Co. F, 26th Va. Cav.

PARSON, HENRY CHESTER: Captain, Co. I, 1st Vt. Cav. U.S.

PATTERSON, JOHN M.: Co. K, 22nd Va. Inf.

PATTERSON, HARVEY: 12th Va. Cav.

PATTON, WILLIAM THOMPSON: Co. A, 28th Va. Inf.

PAXTON, ABNER JERN: Co. K, 22nd Va. Inf.

PAXTON, CYRUS H.: Co. A, 18th Va. Inf.

PAXTON, JACOB A.: Co. F, 23rd Va. Inf.

PAXTON, JOHN S.: Co. A, 18th Va. Inf.

PAXTON, THEOPHILUS: 2nd Lt., Co. F, 23rd Va. Inf.

PAYNE, TUCKER W.: Co. C, 14th Va. Inf.

PEERY, JOHN M.: Co. K, 57th Va. Inf.

PEERY, THOMAS LITTLETON: Co. K, 57th Va. Inf.

PENDLETON, EDMUND: Lt. Col., 3rd La. Bn. Inf., Col., 15th La. Inf.

PENDLETON, WILLIAM W.: Washington Arty. of New Orleans

PERKINS, ANDERSON M.: Danville Arty. & New Market Arty.

PHILLIPS, WILLIAM H.: Co. C, 6th Va. Cav.

PIERSON, WILLIAM FRANKLIN: Captain of Inf. Co. Braxton Co. 1861, 2nd Lt., Co. I, 17th Va. Cav.

PLOGGER, WILLIAM P.: Co. F, 35th Va. Bn. Cav.

PLOTT, ALFRED ABRAHAM: Co. K, 20th Va. Cav.

POAGUE, JOHN E.: Co. K, 22nd Va. Inf.

POINDEXTER, GEORGE: Lt., Baltimore Light Arty. & Moody's La. Arty.

POTEE, THOMAS: 25th Va. Bn. Inf.

POTTER, JACOB M.: Captain Avis' Co., Provost Guard, Staunton.

PRESTON, JOHN THOMAS LEWIS: Lt. Col., 9th Va. Inf., Col. & AAG Gen. Thomas J. Jackson and VMI staff

PRICE, JOHN THOMAS: Co. G, 57th Va. Inf.

PRICE, WILLIAM ANDREW: Purcell Arty.

RAMSEY, JAMES MADISON: Nelson Arty.

RAPP, JOHN H.: Co. I, 28th Va. Inf.

RAWLINGS, BENJAMIN CASON: 1st S.C. Inf. (Gregg's) & Captain, Co. D, 30th Va. Inf.

RAWLINGS, ZACHARIAH HERNDON: Co. A, 30th Va. Inf.

REED, JAMES HOWARD: Marquis' Boys Battery, Staunton

REID, WILLIAM N.: Co. C, 34th Va. Inf.

REILLY, JAMES: Co. H, 2nd Va. Inf.

REVELY, DAVID R.: Co. G, 56th Va. Inf.

RHEA, JOHN SHAW: Co. G, 18th Va. Cav.

RHODES, JOHN JACOB: Captain, Avis' Co., Provost Guard, Staunton

RICHESON, ANDREW JACKSON: Co. C, 43d Bn. Va. Cav.

RISK, JOHN H.: Charlottesville Arty.

RISK, JOHN WILLIAM: Charlottesville Arty.

ROBERTSON, JOHN NEWTON: Co. G, 49th Va. Inf.

ROBINSON, M. S.: Co. K, 22nd Va. Inf.

ROBINSON, STARKEY: Botetourt Arty.

ROBINSON, WILLIAM F.: Co. F, 23rd Va. Inf.

ROBINSON, WILLIAM S.: Co. K, 22nd Va. Inf.

ROGERS, JOSEPH L.: Co. E, 60th Va. Inf.

ROWAN, J. L.: Co I, 2nd Va. Inf.

ROWSEY, WILLIAM H.: Co. H, 49th Va. Inf.

RUFF, JOSEPH SYLVESTER: Co. K, 16th Va. Cav. and QM Dept.

RUFF, DAVID LEWIS: Captain, 22nd Va. Inf. and on staff Gen. "Cerro Gordo" Williams

RULEY, BURTNEY A. WASHINGTON: Co. E, 3rd Conf. Engineers

RULEY, JAMES M.: Co. I, 12th Va. Cav.

RULEY, WILLIAM ALEXANDER: Co. I, 12th Va. Cav.

RUNNELS, SAMUEL: Co. K, 57th Va. Inf.

SALE, JOHN M.: Co. A, 19th Va. Inf.

SALE, ROBERT N.: Co. G, 51st Va. Inf.

SANDEFER, J. C.: Vaughn's Co., 4th Ark. Inf.

SAVILLE, W. A.: Co. E, 22nd Va. Inf.

SCOTT, THOMAS LACKEY: Co. K, 22nd Va. Inf.

SEEBERT, JACOB FRY: Co. K, 22nd Va. Inf.

SEIG, D. A.: Captain Avis' Co., Provost Guard, Staunton

SELLERS, WILLIAM HARVEY: 2nd Lt. & Adj., 5th Tex. Inf., Lt. Col. & AIG, Hood's Division, Col. & AIG Gen., Longstreet's staff.

SEMMES, THOMAS MIDDLETON: Adj., 3rd Ark. Inf., Ordnance Officer, Gen. H. R. Jackson's Brigade, staff, VMI

SHATZER, JOHN P.: Salem Arty.

SHEETS, JAMES WILLIAM: Marquis' Boys Bty., Staunton.

SHERRARD, JOSEPH HOLMES, JR.: Co. F, 2nd Va. Inf.

SHIELDS, JOHN CAMDEN: Captain, Richmond Howitzers, Major, Lt. Colonel and Colonel, C.S.A. Commanded Camp Lee, Richmond, last two years of the war.

SHIPP, SCOTT: Major, 21st Va. Inf. and staff, VMI

SHORTER, JAMES ANDREW: Co. E, 3rd Conf. Engineers

SILVEY, MORGAN: Co. C, 34th Va. Inf.

SIRON, DAVID FRANKLIN: Carpenter's Va. Arty.

SIRON, JOHN M.: 47th Bn. Va. Cav.

SLAGLE, JOHN H.: Co. F, 50th Va. Inf.

SMILEY, WALTER: Carpenter's Va. Arty.

SMITH, FRANCIS WILLIAMSON: Military, Sec. Gen. R. E. Lee, Major, 41st Va. Inf., Major of Arty., Lt. Colonel C.S.A.

SMITH, GEORGE R.: Marquis' Boys Bty., Staunton.

SMITH, HEZEKIAH J.: Co. H, 18th Va. Cav.

SMITH, JAMES: Co. K, 60th Va. Inf.

SMITH, JOHN N.: Co. G, 20th Va. Inf.

SMITH, ROBERT McBRIDE: Co. B, 17th Va. Cav. & Co. B, 39th Bn. Va. Cav.

SMITH, SAMUEL B.: Carpenter's Va. Arty.

SMITH, THOMAS HENDERSON: Adj., 9th Va. Inf., and staff Gen. Fitzhugh Lee

SMITH, THOMAS P.: Co. E, 26th Va. Cav.

SNEAD, WILLIAM JAMES: Co. K, 19th Va. Inf.

SNYDER, WILLIAM S.: Marquis' Boys Bty., Staunton.

STANDARD, PHILIP BEVERLY: Captain, Thomas Va. Arty., Major & Ord. Off. Gen. A. P. Hill's Corps

STANLEY, CHARLES: 7th Va. Cav.

STATON, ANDREW A.: Captain Joel Campbell's Va. Arty.

STATON, ANDREW A.: Captain Watts Co., Col. Booker's Regt. Va. Reserves

STATON, GEORGE W.: Co. F, 50th Va. Inf.

STERRETT, JAMES REID: Captain Avis' Co., Provost Guard, Staunton.

STRAIN, HUGH SCOTT: Surgeon, 110th Ohio Inf. U.S.

STRICKLER, JOHN MONTGOMERY: Marquis' Boys Bty., Staunton

STROTHER, MORDECAI W.: Captain, Co. C, 4th Va. Cav.

SUDDATH, JAMES LITTLETON: 36th Bn. Va. Cav.

SUPINGER, MICHAEL MOCK: Captain Dabney's Va. Arty.

SUTLER, JOHN T.: Co. B, 19th Va. Inf.

SWANN, PORTERFIELD: Dance's Va. Arty.

SWITZER, ABRAHAM: Salem Arty.

SWINK, JAMES V.: Co. B, 8th Va. Cav.

SWINK, ZACHARIAH: Co. G, 31st Va. Inf.

SWISHER, DANIEL Z.: Marquis' Boys Bty., Staunton.

SYERS, DAVID: Carpenter's Va. Arty.

TALLEY, JOHN WINN: Co. G, 3rd Va. Cav.

TAYLOR, ROBERT JOHNSTON: Chaplain, 9th Va. Inf., and Lexington Hospital

TAYLOR, THOMAS DELAWARE: Co. I, 49th Va. Inf.

TAYLOR, WILLIAM: Letcher Arty.

TAYLOR, WILLIAM M.: Clark's Va. Arty.

THOMAS, ARCHIBALD: Co. C, 34th Va. Inf.

THOMPSON, JAMES H., JR.: Captain Avis' Co., Provost Guard, Staunton

TOLLEY, JOHN HENRY: Co. C, 34th Va. Inf.

TOLLEY, JOHN V.: Co. G, 3rd Va. Cav.

TOLLEY, WILLIAM H.: Co. K, 1st Tenn. Cav. (Carter's)

TOOMAN, JAMES: Letcher Arty.

TREVEY, JACOB C.: Co. F, 12th Va. Cav.

TRUSLOW, LEVI: Co. G, 61st Va. Inf.

TUCKER, WHITING D.: C.S.N.

TURNER, THOMAS H.: Co. C, 34th Va. Inf.

TUTTLE, ROMULUS MORRIS: Captain, Co. F, 26th N.C. Inf.

TUTWILLER, ELI SHORES: Captain, 1st Va. Bn. Inf. (Irish) & QM, Lexington

TYLER, SAMUEL P.: Co. D, 19th Va. Bn. Arty.

TYREE, RICHARD C.: Kirkpatrick's Arty.

UNROE, ADAM: Co. H, 18th Va. Cav.

VEST, DAVID J.: Co. I, 20th Va. Cav.

VEST, GEORGE W.: Captain Avis' Co., Provost Guard, Staunton

VEST, JAMES M.: Co. I, 20th Va. Cav.

VEST, SAMUEL G.: Co. I, 20th Va. Cav.

VIA, HARRISON: Co. K, 56th Va. Inf.

VIA, MADISON: Co. A, 17th Va. Inf.

VIA, WILLIAM M.: Marquis' Boys Bty., Staunton.

WADDELL, ADDISON ALEXANDER: Charlottesville Arty.

WADDELL, JAMES HENRY: Adj., 26th Va. Inf.

WADDELL, WILLIAM HENRY HARRISON: Co. A, 62nd Va. Inf.

WADE, ALGERNON SIDNEY: Co. B, 12th Va. Cav.

WADE, BENJAMIN FRANKLIN: Co. C, 19th Va. Inf.

WADE, HORACE MORRALL: Co. I, 12th Va. Cav.

WALKER, GEORGE M.: Carpenter's Va. Arty.

WALKER, GEORGE S.: Asst. Surgeon, 21st N.C. Inf.

WALKER, JAMES STEEL: Carpenter's Va. Arty.

WALKER, JAMES T.: Carpenter's Va. Arty.

WALKER, JOHN: Wheat's La. Bn. Inf.

WALKER, JOHN W.: Carpenter's Va. Arty.

WALKER, JOHN WILLIAM: Co. B, 12th Va. Cav.

WALKER, SAMUEL S.: Carpenter's Va. Arty.

WALKER, THOMAS JEFFERSON: Co. D, 12th Va. Cav.

WALKER, WILLIAM THOMAS: 2nd Ga. Bn. Inf.

WALKER, ZECHARIAH JOHNSTON: Asst. Surgeon, 17th Va. Cav. & Surgeon McCausland's Brigade

WALKUP, JOSEPH WALKER: Chaplain, 9th Va. Inf.

WALLACE, ALEXANDER AUGUSTUS: 22nd Va. Cav.

WALLACE, WILLIAM A.: Bryan's Va. Arty.

WATKINS, JOHN HASTINGS: Co. B, 12th Va. Cav.

WATSON, LEWIS L.: 58th Va. Inf. & Co. K, 28th Va. Inf.

WATTS, THOMAS NELSON: Kirkpatrick's Va. Arty.

WAYLAND, WILLIAM R.: Co. K, 19th Va. Inf.

WELCH, JOHN H.: Lt., Co. I, 31st Indiana Inf. U.S.

WELCHER, JOSEPH: Co. F, 60th Va. Inf.

WEST, JOHN W.: Co. E, 7th Va. Cav.

WHEELER, ESAU: Captain Hall's Co., 50th Va. Inf.

WHITE, ARTHUR: Co. E, 2nd Va. Cav.

WHITE, EMORY ASBURY DULIN: Co. A, 1st Va. Cav. & McNeill's Rangers

WHITE, HENRY MILTON: Chaplain, Richmond Howitzers & Hardaway's Bn. Arty.

WHITESIDES, CHARLES H.: Co. G, 51st Va. Inf.

WILKERSON, WILLIAM V.: Co. B, 10th Va. Bn. Arty.

WILLIAMS, ARCHIE P.: Co. B, 2nd La. Inf., Lt., ADC Gen. A. P. Bagby, Major & IG.

WILLIAMSON, HENRY WATSON: Lt. Col., 6th Va. Inf. & staff, VMI

WILSON, GEORGE W.: Salem Arty.

WILSON, J. W.: Co. H, 16th Va. Cav.

WILSON, JAMES: 12th Va. Cav.

WILSON, JAMES A.: Wickham's Brigade Va. Cav.

WILSON, JAMES C.: 10th Va. Cav.

WILSON, JOHN: Co. H, 33rd Va. Inf.

WILSON, WILLIAM LYNN: Co. B, 12th Va. Cav.

WILSON, WILLIAM THOMAS: Lt., Co. A, 18th Va. Inf.

WINN, WILLIAM J.: Carpenter's Va. Arty.

WISEMAN, JACOB A.: Marquis' Boys Bty., Staunton.

WOODWARD, WILLIAM A.: Co. C, 19th Va. Inf.

WOOLFOLK, JOHN SAUNDERS: Co. E, 34th Va. Inf.

WORLEY, ROBERT W.: Lee Arty.

WORLEY, WILLIAM J.: Lee Arty.

WRIGHT, AUGUSTUS TYLER: Co. G, 18th Va. Cav.

WRIGHT, JOHN D.: Kirkpatrick's Arty.

WRIGHT, LEWIS B.: Co. F, 28th Va. Inf.

YOUELL, WILLIAM M.: Co. H, 33rd Va. Inf.

YOUNT, SAMUEL: Co. D, 19th Va. Inf. & Co. G, 42nd Va. Inf.

ZIMMERMAN, GEORGE P.: Co. C, 2nd Va. Cav.

ZOLLMAN, JAMES MADISON: Dabney's Va. Arty.

CIVIL WAR VETERANS OF ROCKBRIDGE COUNTY — UNITS UNKNOWN

ALPHIN, THOMAS
ARGENBRIGHT, J. FRANK
BARCLAY, JAMES PAXTON
BEARD, PHILIP: Cavalry
BISHOP, HENRY D.
BONES, BUTTON GWINNETT
BOSWORTH, ELAM BUCKLEY
BOWYER, MICHAEL M.
BROCKENBROUGH, JOHN WHITE
BROWN, WILLIAM SITTLINGTON: Colonel
BRYANT, CHARLES A.
BYARS, RICHARD
CAMPBELL, WILLIAM
CARTER, WILLIAM L.
CARUTHERS, WILLIAM H.
CHESTER, WILLIAM YEARMANS: Captain
CHITTUM, JAMES FLETCHER
CHITTUM, ZACHARIAH ALEXANDER
CLARKE, JOHN
CLEVELAND, J. P.: Pickett's Div.
COLE, J. W.
COLEMAN, W. A.: Col. Pickett's Regt.
COOK, JOHN P.: QM Dept., Lynchburg
CORMIER, T. W.: Stonewall Brig.
DAVIS, T. C.
DEACON, JOHN J.
DICKINSON, JOHN BELL
DODD, SAMUEL M.: Surgeon
DONALD, WILLIAM H.
DUNLAP, JOHN M.
DUVALL, ELI: Captain
ECHOLS, ROBERT JOSEPH: Major
ENGLEMAN, PHILIP
ESTILL, HENRY MILLER: Lt., Ord. Dept.
FITZPATRICK, I. W.
FITZPATRICK, R. M. C.
FORBES, JOSEPH N.
FULWIDER, JOHN
GIBBS, GEORGE S.: Al. Cadets
GIBBS, JOHN TRACY: QM, U. Ala.
GLASGOW, FRANCIS THOMAS: Manager, Tredegar Ironworks, Richmond
GRAHAM, JAMES
HALL, ALEXANDER: Treas. Dept.
HALL, JOHN D.
HALL, RUFUS
HALL, WILLIAM S.
HAMILTON, ISAAC MONTGOMERY
HAMILTON, JOHN LETCHER
HARRIS, JOHN HENRY: Chaplain
HARRIS, ROBERT JACKSON: Cavalry
HARTLESS, CHARLES HUNTER
HARTSOOK, WILLIAM S.
HAYS, JAMES H.: Comm. Dept., Staunton & Lexington
HERRINGTON, E. E.: U.S.N.
HITE, ROBERT A.
HOLT, PLEASANT
HOSTETTER, THOMAS: Guard, Jordan's Furnace
HOUSTON, JOHN WILSON: Lt., Inf.
HUFFMAN, GEORGE H.
HUMPHREYS, WILLIAM F.: Surgeon
ICENHOWER, ANDREW J.
JACCHERI, POMPEO: Courier, Pickett's Div.
JOHNSON, EDWARD: General
JOHNSTON, WILLIAM PRESTON: Lt. Col., Pres. Davis' staff
JONES, WILLIAM DAVID: Surgeon
JORDAN, JOHN F.
JUNKIN, EBENEZER DICKEY: Chaplain
KENDALL, WILLIAM H.: U.S. Army
KENNEY, JAMES MADISON
KNICK, HUGH THOMAS
LACY, BEVERLY TUCKER: Chaplain

ANDERSON, C. T.
AYLOR, R. S.
BARTON, HOWARD THORNTON: Surgeon
BEARD, WILLIAM
BIBB, W. C.
BONES, HURLY J.
BOWLING, JOHN J.: Nitre & Mining B.
BRADDS, JAMES M.
BROOKE, JOHN MERCER: Colonel CSN
BROWN, WILLIAM J.
BURGER, PETER N.: Cavalry
CAMPBELL, SAMUEL R.: Surgeon
CARROLL, J. F.
CARUTHERS, SAMUEL: Surgeon
CASH, ROBERT R.
CHITTUM, JAMES ALEXANDER
CHITTUM, ROBERT G.
CLARK, A. DAVIDSON
CLAYBORN, JOHN L.
CLYMER, J. W.
COLEMAN, JOHN WILLIAM
COLSTON, EDWARD
COOPER, WINSTON
CUMMINGS, JOHN D.
DAY, B. F.: U.S. Army
DEDERICK, JACOB
DILLON, THOMAS HAILE
DOLD, SAMUEL MILLER: Surgeon
DOUGLASS, EDWARD W.: Comm. Dept.
DOYE, WILLIAM
EAST, SAMUEL A.
ELHART, ADOPH: Major & Paymaster
ESTILL, CHARLES PATRICK: Captain
FAUBER, JOSEPH M.: QM Dept.
FITZPATRICK, JAMES B.: Chaplain
FLINT, JAMES
FULTON, WILLIAM H.
GIBBS, CHARLES W.: Courier
GIBBS, JAMES EDWARD ALLEN: 1st Lt. Ord Dept.
GILMORE, PAXTON
GLASGOW, WILLIAM ANDERSON
GORDON, EDWARD CLIFFORD: Lt., Ord. Dept.
GREEN, WILLIAM H.
HALL, GEORGE P.
HALL, JOHN F.
HALL, SHANNON H.
HAMILTON, ALEXANDER L.: Chaplain
HAMILTON, JACOB
HARLOW, ROBERT N.
HARRIS, JOHN W.
HART, DAVID HENRY: QM Dept., Staunton
HARTSOOK, JAMES W.
HAWES, MILO C.
HENDERSON, FRANCIS WILLIAM: Capt.
HIGGINS, GEORGE W.
HOLBROOK, JOHN H.: QM Dept.
HORNER, DAVID B.
HOUSTON, CHARLES
HOUSTON, MATTHEW HALE: Surgeon
HUMES, JAMES EDWIN: Lt.
HUCHESON, MATTHEW W.
IRVINE, ROBERT W.
JOHNSON, WILLIAM H.: Capt. Joseph R. Anderson's Co.
JONES, GEORGE W.
JORDAN, BENJAMIN J.: Colonel
JUDY, JOHN
KELLY, J. E.
KENNEDY, WILLIAM RICE
KNIGHTON, JOHN T.
KOONTZ, ALBERT L.: U.S. Navy
LAM, JOHN W.
LEECH, ANDREW FRANKLIN

LANE, FRANCIS
LEECH, HENRY MILLER
LEITCH, JAMES LEWIS: Capt., Ord. Dept.
LEYBURN, GEORGE WILLIAM: Chaplain
LITTLE, JOHN W.
LOGAN, JOSEPH PAYNE: Surgeon
LOWMAN, WILLIAM C.
LUCAS, JOHN
MACKEY, NATHAN A.
MAHONEY, TIFFORD P.: Arty.
MAY, WILLIAM DAVID
MILLER, HENRY CLAY
MOHLER, ABRAHAM NEWKIRK WEAVER
MOORE, ALEXANDER ARCHER
MOORE, JOHN
MORRIS, WASHINGTON FLOYD
MORRISON, ROBERT HALL
MORRISON, JOSIAH
MULLEN, WILLIAM
McCARTHY, WILLIAM H.
McCLURER, JOHN EDMONDSON: Surgeon
McCORKLE, WILLIAM: Major on Gen. Price's staff
McDONALD, MARSHALL M.: Capt. of Engineers, Ord. Officer, Vicksburg, IG, Gen. J. B. Johnston's staff, B. Gen. Engineers
McFADDEN, ANDREW
McLEOD, ROBERT TILDEN: Cavalry
McMILLAN, CHARLES NEWTON: Major, Instructor of Cavalry, Richmond
NEWMAN, WILSON: Lt.
NUCKOLS, G. H.
ORBISON, W. C.: Captain
PARKER, CHARLES W.
PATTON, TRICE
PAXTON, ANDREW JACKSON: Major
PAXTON, JAMES F.
PAXTON, JAMES J.: Stonewall Brig.
PAXTON, NATHANIEL JOSEPH
PAXTON, THOMAS: Colonel, Stonewall Brigade
PAXTON, WILLIAM HENRY: Hays La. Brigade
PENDLETON, ALEXANDER SWIFT: Lt. Col. & AAG Gen. T. J. Jackson's staff and Gen. Ewell & Early's staffs
POINTER, CHARLES H.
POTTER, JAMES J., JR.
PRITCHARD, N. B.
RAMSEY, WILLIAM
REED, ALEXANDER S.: Stonewall Brigade
REID, ROBERT E.
RHODES, G. W.: U.S. Army
RHODES, JOHN PEYTON
ROADCAP, ABRAHAM B.
ROBERTSON, THOMAS S.
ROGERS, JOHN DALRYMPLE: Major & QM Gen. N.G. Evans' Brigade
RUFF, WILLIAM ALEXANDER: Comm. Dept.
SAMUELS, ROBERT G.
SCOTT, JOHN ANDREW: Chaplain
SHELTON, ______
SHORTER, JAMES A.
SHOULDER, F. W.
SMITH, CHARLES VIRGINIUS
SNIDER, ELI
STANLEY, WILLIAM
STEELE, DAVID W.
STONER, JAMES W.: QM Dept.
STRAIN, EUSEBIUS HENRY: Surgeon
STRICKLER, MOSES: QM Dept., Loring's Div.
SWISHER, WILLIAM: Infantry
TEMPLETON, BENJAMIN PORTER
TERRY, WILLIAM
THOMPSON, CHARLES S.: Imboden's Brig.
THORNTON, ERASTUS S.
TOMBLIN, JOHN L.: U.S. Army
TOOMAN, JOHN T. F.
TUTWILER, THOMAS H.: QM, Lexington
UNROE, JOHN TRIBBETT: Captain, Pickett's Div.
VEST, ALEXANDER: Texas Regt.

LEECH, SAMUEL SCOTT
LEWIS, JAMES S. B.
LINDSAY, WILLIAM: Captain, Ky. unit
LOCHER, CHARLES HESS: QM Dept., Lynchburg
LOWEN, JOHN W.
LUCAS, JAMES
LUCK, SIDNEY JOSEPH
MAHONEY, CHARLES W.: Arty.
MANSPILE, RENARD KIMBERLINE: Cav.
MAYS, ELI
MILLER, JAME M.: CSA & USA
MILLER, JOHN A.
MILLER, JOHN ADDISON
MONTGOMERY, S. M.: Wharton's Div.
MOORE, J. E.: Captain of Arty.
MOORE, WILLIAM
MORRISON, JAMES HORACE: Lt. & ADC Gen. Pemberton, Colonel, U. Ala.
MOSELEY, GEORGE W.
MULLEY, WILLIAM
McCLUNG, JOSEPH: Surgeon
McCORKLE, ALFRED LEYBURN: Surgeon
McCORMICK, GEORGE B.
McDOWELL, JAMES E.: Surgeon
McELWEE, WILLIAM MEEK: Chaplin
McFADDEN, ABRAHAM: Greenbrier Rifles
McFADDEN, HUGH
McMANAMA, JOHN A.: Teamster
McNUTT, BENJAMIN GRIGSBY
McNUTT, JAMES MORTON
NORTHERN, WILLIAM H.
NUCKOLS, WILLIAM MARSHALL: Courier
PAINTER, JACOB M.
PATTON, ALFRED TAYLOR
PAXTON, ALEXANDER McNUTT: Major & QM
PAXTON, JAMES C.: Major & QM
PAXTON, JAMES GARDINER: Major
PAXTON, JOHN HENRY
PAXTON, SAMUEL: Major
PAXTON, WILLIAM GALLATIN: Major, Miss. Cav.
PAXTON, WILLIAM W.: Stonewall Brigade
PERRY, JOHN THEODORE SPRIGGS: Colonel
PETTIGREW, GEORGE WASHINGTON: Capt., Comm. Dept.
POTTER, GEORGE T.
PRESTON, THOMAS LEWIS: Chaplain
RAMSEY, THOMAS
RAY, HENRY B.
REID, HENRY J.
REID, WILLIAM C.
RHODES, JOHN JAMES
RICE, THOMAS J.
ROBERTSON, P. LEWIS: Mason's Pioneers
RODRICK, JACOB H.: Md. Co., Stonewall Brigade, courier, Gen. T. J. Jackson.
RUCKER, DANIEL H.: U.S. Army
RUSSELL, JAMES P.: QM Dept.
SAVILLE, WILLIAM
SHAFER, FRANK
SHORT, ROBERT PAXTON
SHORTER, JOHN THOMAS
SIPE, WILLIAM A.
SMITH, ROBERT G.: Nitre Mining Bureau
SNIDER, WILLIAM FRANKLIN
STATON, ANDREW T.
STEVENS, WALTER LE CONTE: S.C. unit
STRAIN, DAVID ELDRED: Surgeon
STRICKLER, CORNELIUS HERRON
SULLIVAN, JOHN: Cavalry
TEAFORD, DAVID
TERRY, MARSHALL S.
TETER, WILLIAM L.: U.S. Army
THOMPSON, PARIS
TOLLEY, FRANCIS MARION
TOLLEY, WILLIAM LARKIN TAYLOR: U.S. Army
TRICE, GEORGE W., JR.
UNROE, ALBERT W.
VANDERSLICE, GEORGE CURTIS

WADDY, J. J.: Captain
WALKER, SAMUEL AUGUSTUS: Surgeon
WALLACE, GEORGE PRESTON
WALTON, WILLIAM W.
WATTS, JAMES NELSON: Bugler
WELCH, ELISHA GRIGSBY
WHITE, JAMES M.
WHITESELL, A. H.
WHITSELL, JEREMIAH
WILHELM, RUFUS
WILKINSON, JAMES LEONIDUS
WILKINSON, WILLIAM WESLEY: Cavalry
WILLS, FLEMING SPOTTSWOOD
WILSON, ROBERT LUCIAN
WISEMAN, JOSEPH H.
WRIGHT, GEORGE W.

WADDELL, JAMES ALEXANDER: Surgeon
WALKER, JAMES A.
WALKER, SAMUEL BRANCH
WALTON, THOMAS H.
WARD, JAMES R.: Col. Pickett's Regt., Ala.
WELCH, ALEXANDER
WELCH, JOHN W.: Inf.
WHITE, JOHN
WHITESELL, GEORGE H.
WILCHER, THOMAS H.: U.S. Army
WILLIAMS, A. G.: Captain, Inf.
WILSON, BENJAMIN F.
WILSON, WILLIAM F.: Colonel
WRIGHT, CARY
WRIGHT, JOHN R. F.

VIRGINIA MILITARY INSTITUTE GRADUATES AND CADETS FROM ROCKBRIDGE COUNTY AND STAFF DURING THE WAR NOT OTHERWISE IDENTIFIED

ANDERSON, FRANCIS THOMAS, JR.: Cadet, served in trenches around Richmond.

BLAIR, WILLIAM BARRETT: Colonel, Commissary General of Virginia 1861-65.

BLUM, JOHN ALEXANDER: Cadet. Present McDowell Campaign & Signal Corps.

BOWIE, WALTER, JR.: 11th Va. Inf., Co.G, 40th Va. Inf., Staff, VMI, Co. C, 9th Va. Cav., CSN, 43rd Bn. Va. Cav.

BOWYER, WILLIAM McDONALD: Cadet. Served in trenches around Richmond.

BROCKENBROUGH, ROBERT LEWIS: Cadet. New Market Bn. 1st Lt. in General Joseph E. Johnston's Army.

BULL, WILLIAM ROOT: Private Secretary, Steward & Commissary, VMI 1862-65.

CARMICHAEL, JOHN: Cadet. New Market Bn. and staff officer in Ga.

CATLETT, RICHARD HENRY: Treasurer. Lt. Col. & AAG Gen. Echols staff, ADC Gov. Letcher, AAG, Gen. Kemper.

COLONNA, BENJAMIN AZARIAH: Cadet. New Market Bn., Marquis' Boys Battery, Staunton, 21st and 31st Va. Inf., 1st Foreign Bn.

CROCKEN, JAMES HENRY: Musician.

ECHOLS, JOSEPH ROWLAND: Cadet. New Market Bn. Served in trenches around Richmond.

ESTILL, HENRY MILLER: Surgeon, 51st Va. Inf. and VMI.

EVANS, JOHN F.: Musician. Also served in Ord. Dept., Richmond.

GILHAM, WILLIAM HENRY: Cadet. Served in trenches around Richmond.

HAMPSEY, JOHN: Ordnance Sergeant 1861-1865.

JOHNSON, PORTER: Cadet. New Market Bn., Lt., 8th Conf. Foreign Bn.

LETCHER, SAMUEL HOUSTON: Cadet. New Market Bn., 2nd Lt., CSA.

MADISON, ROBERT LEWIS: Surgeon.

MARKS, JACOB: Musician, New Market Btn.

McCARTHY, JEREMIAH: Employee.

McCORKLE, JAMES WILLIAM: Cadet. New Market Bn.

MOHLER, DAVID GUIN: Cadet. New Market Bn. and 43rd Bn. Va. Cav.

PRESTON, EDMUND RANDOLPH: Cadet. McDowell Campaign.

PRESTON, FRANK C.: Captain, Co. B, New Market Bn.

SHIELDS, JOHN HARDY: Cadet. New Market Bn.

SMITH, FRANCIS HENNY: B. Gen. Superintendent. Colonel, 19th Va. Inf. 1861-62.

SMITH, FRANCIS HENNY, JR.: Cadet. Served in trenches around Richmond.

SMITH, ISAAC WILLIAMS: Capt. of Pontoon Service, Colonel of Engineers.

STAPLES, WILLIAM RICHARD: Drummer, New Market Bn.

TRUEHEART, DANIEL, JR.: Served at Pensacola 1861. Major and Chief of Artillery, Gen. Thomas J. Jackson's staff.

TUTWILER, EDWARD MAGRUDER: Cadet. New Market Bn. CSA.

WHITE, JOHN SPROUL: Cadet. New Market Bn. CSA.

WHITE, ROBERT JOSEPH: Cadet. New Market Bn. Captured the flag of the 27th N.Y. Inf.

WHITWELL, JOHN COYLE: Captain & Commissary Officer. New Market Bn.

WILLIAMSON, THOMAS HOOMES: Major of Engineers on Gen. T. J. Jackson's staff, Lt. Col. of Engineers on staff Gen. Holmes. Captain of Patrol, Lexington, 12/62 and Captain of Home Guard Co., Rockbridge Co. 1863, while on staff of VMI.

WINGFIELD, THOMAS SMITH: Musician. Drillmaster, Transportation Agent, CSN. Manufacturer of explosives, Charlotte, N.C.

FOOTNOTES

The footnotes do not include all of the material consulted in the preparation of this book. All direct quotations from unpublished sources are cited. Most titles have been shortened for brevity after their first usage. The complete citations are contained in the bibliography. *The Lexington Gazette and General Advertiser* underwent several name changes over the years but is cited as *Gazette* throughout. The most frequently used references are indicated by the author's surname. Where there is more than one title by an author, a short title is used. Since there are two Wise's who wrote on this period, *The Military History of the Virginia Military Institute from 1839-1865* has been shortened to *The Military History of V.M.I.* Where there is more than one title by the same author, the last name of the person written on is used. An example is Charles W. Turner, editor, *Lieutenant John Newton Lyle, the Career of the Liberty Hall Volunteers,* is listed as *Lyle.* Diaries and letters are listed by the actual writer's surname after the first usage. *The War of the Rebellion: A Compilation of the Official Records of the Union and Confederate Armies,* Series I-VI, 128 volumes, 1880-1901, is called *O.R.* One exception is the James Dorman Davidson correspondence. Because the Davidson material is located in at least three repositories, the letters published in the Civil War History is identified as *Life Behind Confederate Lines,* after the first usage.

Chapter I

1. Elizabeth Randolph Preston Allan. *A March Past: Reminiscences of Elizabeth Randolph Preston Allan,* Janet Allan Bryant, ed., Richmond: The Dietz Press, 1938, pp. 101-109.
2. F. N. Boney, *John Letcher of Virginia,* University, Ala.: University of Ala. Press, 1966, pp. 74-88.
3. *Ibid.,* pp. 98-100. *Lexington Gazette and General Advertiser,* 20 January 1927. For additional beliefs on the causes of the war see Rev. H. M. White, ed. *Reverend William S. White, D.D., and His Times,* Richmond: Presbyterian Committee of Publication, 1891.
4. *Valley Star,* Lexington, Va. 13 December 1860. Bell received 1201 votes, Breckinridge 344 and Douglas 52. Boney, pp. 100-102.
5. Minutes of the Lexington Town Council, 30 November 1860. *Gazette,* 31 January 1861. Allan, pp. xxviii-xxx and p. 111. For an excellent description of a pre-war Christmas in Lexington see Allan, pp. 109-111. *Star,* 10 January 1861.
6. Boney, pp. 104-105.
7. Boatner, Mark M., III. *The Civil War Dictionary,* New York: David McKay Co., Inc., 1966, pp. 729-730. Allan, p. 111. *Gazette,* 20 January 1927. Moore and Dorman each received about 1,869 votes, Brockenbrough 293 and Baldwin 72.
8. *Gazette,* 20 January 1927. *Star,* 14 March 1861. Boney, pp. 106-110.
9. Washington College Faculty Minutes, 26 March and 1 April 1861.

Washington and Lee University Historical Papers, No. 6, 1904, Lynchburg: J. P. Bell Co., Printers, 1904, pp. 124-125. William W. Greenlee, *All the Year Round,* Charles Dickens, ed., London, England, 17 May 1862, William W. Greenlee Diary, 4-5 March 1861.

10. Boney, pp. 111-112. *Gazette,* 20 January 1927.
11. Boney, p. 112.
12. *Star,* 18 April 1861. Greenlee Diary, 13 April 1861.
13. A. C. S. Gatewood letter to "My Dear Parents," 15 April 1861, A. C. S. Greenwood File, V.M.I. Archives.
14. Alexander S. Paxton, *Memory Days,* New York and Washington: The Neale Publishing Co., 1908, pp. 229-233. Greenlee Diary, 14 April 1861. James H. Wood, *The War, Stonewall Jackson: his Campaigns and Battles, the Regiment as I Saw Them,* Cumberland, Md.: The Eddy Press, 1916, pp. 5-12. R. A. Marr, Jr., *Echos of the Virginia Military Institute,* J. E. Johnson, ed., Lexington: 1937, p. 14. Jennings C. Wise, *The Military History of the Virginia Military Institute from 1839-1865,* Lynchburg: J. P. Bell Co., Inc., 1915, pp. 126-133.
15. Charles W. Turner, ed., *Lieutenant John Newton Lyle: The Career of the Liberty Hall Volunteers,* Verona, Va.: The McClure Press, 1906, p. 4.
16. Allan, pp. 117-118.
17. *Lyle,* p. 4. Greenlee Diary, 5 March 1861.
18. W&LU Papers, pp. 124-125. Washington College Faculty Minutes, 17 April 1861. Greenlee Diary, 17 April 1861. George Junkin, "Exodus of Dr. Junkin," *Presbyterian Standard* (17 May 1861), in *David X. Junkin, D.D., LL.D., A Historical Biography,* Philadelphia: 1871.
19. Boney, pp. 112-113.

Chapter II

1. White, pp. 171-173.
2. Bruce S. Greenwalt, ed., "Life Behind Confederate Lines in Virginia, The Correspondence of James Dorman Davidson," *Civil War History,* Vol. 16, September 1970, pp. 207-209.
3. H. G. Davidson letter to "Dear Brother" 18 April 1861: J. H. Davidson Collection, Indiana State University, Indianapolis, Ind.
4. A. C. S. Gatewood letter to "My Dear, Parents" 18 April 1861, V.M.I. Archives. Greenlee Diary, 18 April 1861. *Military History of V.M.I.,* p. 138.
5. Thomas A. Stevenson letter to "My Dear Sister" 19 April 1861, V.M.I. Archives.
6. *Star,* 21 April 1861. *Gazette,* 26 April and 23 May 1861.
7. Allan, pp. 127-129. *Military History of VMI,* pp. 139-142. Francis H. Smith, *History of the Virginia Military Institute It's Building and Rebuilding,* Lynchburg: J. P. Bell Co., Inc., 1912, p. 181. Thomas W. Davis, ed., *A Crowd of Honorable Youths,* Lexington, Va.: V.M.I. Sesquicentennial Committee, 1988, pp. 36-37.
8. *Washington College Record,* pp. 124-125. *Military History of V.M.I.,* p. 139. Andrew Brooks letter to "Dear Mattie" 6 May 1861, Reid Family Papers, W&LU.
9. *Lyle,* pp. 6-9. Susan P. Lee, *Memoirs of William Nelson Pendleton, D.D.,* Philadelphia: J. P. Lippencott Co., 1893, p. 137.
10. *Star,* 20 April 1861. *Gazette,* 25 April 1861.

11. Minutes of the Lexington Town Council, 20 April 1861. Washington College Faculty Minutes, 29 April 1861.
12. Charles W. Turner, ed., *The Diary of Henry Boswell Jones of Brownsburg, 1842-1871,* Verona, Va.: McClure Press, 1979, 27 April 1861. James D. Davidson letter to "Dear Alexander" 24 May 1861, Davidson Family Papers, W&LU.
13. Thomas McGuffin letter to John B. McGuffin 22 April 1861, McGuffin Family Papers, UVa.
14. Minute Book of Rockbridge County, May 1861.
15. *Lyle,* pp. 7-8.
16. Thomas McGuffin letter to John B. McGuffin 7 May 1861, McGuffin Family Papers, UVa.
17. *Star,* 2 and 9 May 1861. *Gazette,* 23 May 1861. Lee, pp. 137-138.
18. William G. Bean, ed., "A House Divided: The Civil War Letters of a Virginia Family," *The Virginia Magazine of History and Biography,* Vol. 59, October, 1951, pp. 398-400.
19. *Star,* 16 May 1861. *Gazette,* 23 May 1861. Robert J. Driver, Jr., *The 1st and 2nd Rockbridge Artillery,* Lynchburg: H. E. Howard, Inc., 1987, p. 2. Lee, p. 140.
20. *Star,* 30 May 1861.
21. Boney, p. 116.
22. M. S. Roadcap, letter to "My Son." 8 June 1861. McGuffin Family Papers, U.Va.
23. *Star,* 6 June 1861. *Gazette,* 20 January 1927.
24. *Lyle,* pp. 10-11. Paxton, pp. 275-276. *Jones Diary,* 8 June 1861. William S. White, *Sketches of the Life of Captain Hugh A. White, of the Stonewall Brigade,* Columbia, S.C.: 1864, pp. 46-48.
25. *Star,* 13 June 1861.
26. Allan, pp. 127-129.
27. *Gazette,* 13 June 1861.
28. *Lyle,* p. 9. *Gazette,* 13 June 1861. *Washington College Records,* p. 126. "Life Behind Confederate Lines in Virginia," pp. 214-215.
29. *Gazette,* 13 June 1861.
30. *Ibid.,* 4 June and 18 July 1861. Henry Boley, *Lexington in Old Virginia,* Richmond: Garrett & Massie Publishers, 1935, p. 100. A. C. S. Gatewood letter to "My Dear Parents" 29 May 1861, V.M.I. Archives.
31. *Gazette,* 20 June and 18 July 1861. Davis, p. 37.
32. *Gazette,* 27 June and 4 July 1861.
33. *Ibid.,* 18 July 1861. Robert J. Driver, Jr., *14th Virginia Cavalry,* Lynchburg: H. E. Howard, Inc., 1988, p. 8.
34. *Gazette,* 25 July, 1 August and 8 August 1861. Oren F. Morton, *A History of Rockbridge County, Virginia,* Baltimore: Regional Publishing Co., 1980, pp. 126-127. *1st and 2nd Rockbridge Artillery,* p. 97. Robert J. Driver, Jr., *52nd Virginia Infantry,* Lynchburg: H. E. Howard, Inc., 1986, pp. 1-2. Lee A. Wallace, Jr., *A Guide to Virginia Military Organizations 1861-1865,* 2nd edition, Lynchburg: H. E. Howard, Inc., 1986, p. 136.
35. *Gazette,* 25 July 1861. Boney, p. 139.
36. *Gazette,* 8 August 1861. William A. Albaugh III and Edward N. Simmons, *Confederate Arms,* New York: Bonanza Books, 1957, p. 228.
37. John D. Capers, "Virginia Iron Furnaces of the Confederacy," *Virginia Cavalcade,* August 1967, pp. 10-16. Charles B. Dew, *Ironmaker to the Con-*

federacy, Joseph R. Anderson and the Tredegar Ironworks, New Haven: Yale University Press, 1966, p. 100. Records of the Monmouth Cloth Factory 1861-1865, Rockbridge Historical Society. Boley, pp. 100-101. Journal of Colonel J. T. L. Preston July-September 1861. V.M.I. Archives, 20 September 1861.

38. Francis H. Smith letter to James C. Bruce 10 September 1861. Document No. 11, V.M.I. Board of Visitors Records 1861 and 19 December 1861. Albaugh, p. 239.
39. *Gazette,* 2, 15 and 23 August 1861.
40. *Ibid.*, 5 and 17 September 1861.
41. *Ibid.*
42. *Ibid.*, 19 September, 17 October and 17 November 1861.
43. *Ibid.*, 24 and 31 October and 7 November 1861.
44. *Ibid.*, 21 and 28 November and 5 December 1861. J. William Jones, *Christ In Camp,* Richmond: B. F. Johnson and Co., 1887, pp. 174-175 and 212-214.
45. *Gazette,* 19 and 26 December 1861. *1st and 2nd Rockbridge Artillery,* pp. 101-102. *52nd Virginia Infantry,* pp. 7-8. Andrew Patterson letter to "Dear Brother" 7 January 1862, A. Patterson Papers, UVa. *O.R.*, Vol. 5, pp. 456-468.
46. Benjamin F. Templeton letter "To My Dear Children" undated, Templeton Family Papers, W&LU.
47. Allan, pp. 136-137. James D. Davidson letter to John Letcher 28 December 1861, Letcher Family Papers, W&LU.

Chapter III

1. *Gazette,* 2 January 1862.
2. *Ibid.*, 6, 19 and 27 February 1862.
3. *Ibid.*, 27 February 1862. Boney, pp. 156-157. Minute Book of Rockbridge County, 12 March-4 April 1862.
4. *Gazette,* 27 March, 3 and 17 April 1862. Boney, p. 157.
5. *Gazette,* 10 April and 1 May 1862. Minute Book of Rockbridge County, 5 May 1862. *O.R.*, Vol. 12, pp. 334-410. *1st and 2nd Rockbridge Artillery,* pp. 14-18.
6. *Gazette,* 27 March, 3 and 17 April 1862. Minutes of the Lexington Town Council 7 February 1862.
7. Elizabeth Randolph Preston Allan, *The Life and Letters of Margaret Junkin Preston,* New York: Houghton-Mifflin, 1903, pp. 134-137.
8. *Gazette,* 10 April 1862.
9. *Ibid.*, 1 May 1862. Richard Armstrong, *Ambush at Williamsville,* Staunton: Minuteman Press, 1986.
10. *Gazette,* 8 and 15 May 1862. *Preston,* pp. 138-139. *O.R.*, Vol. 12, pp. 460-491. *52nd Virginia Infantry,* pp. 12-15.
11. *Gazette,* 8 and 29 May 1862.
12. *Ibid.*, 22 May 1862.
13. *Ibid.*, 16 and 29 May 1862. Allan, pp. 144-145. *Preston,* pp. 141-146.
14. *Gazette,* 12 June 1862.
15. *Ibid.*, 12 and 19 June and 7 August 1862.
16. *Ibid.*, 19 June 1862. Washington College Faculty letter to the Surgeon General of the Confederate States 24 July 1862, Washington College Faculty Correspondence, W&LU.

17. *Gazette,* 7 August 1862.
18. *Ibid.*
19. Minute Book of Rockbridge County, 7 July 1862.
20. *Gazette,* 7 August 1862.
21. *Ibid.,* 28 August 4, 11, 18 and 25 September 1862. *Preston,* pp. 146-150. Allan, pp. 146-150.
22. *Gazette,* 14 and 29 August, 1 September and 1 October 1862. H. McC. Davidson letter to "Dear Willy" 5 July 1862, Davidson Family Papers, W&LU
23. *Gazette,* 18 and 25 September 1862.
24. John S. Wise, *The End of An Era,* Boston & New York: Houghton-Mifflin Co., 1899, pp. 232-234.
25. *Gazette,* 9 October 1862 and 3 January 1863.
26. *Ibid.,* 11 September 1862.
27. *Ibid.,* 4 September, 23 October, 6 and 20 November 1862.
28. *Ibid.,* 12 February 1863.
29. *Ibid.,* 20 and 27 November, and 4 December 1862, 8 January 1863.
30. *Ibid.,* 1, 8, 29 January 1963.
31. *Preston,* pp. 157-158.

Chapter IV

1. *Gazette,* 13 and 20 November 1862 and 8 January 1863. Minute Book of Rockbridge County, 5 January 1863. *Preston,* p. 159.
2. Minute Book of Rockbridge County, 5 January 1863.
3. *Ibid.,* 5 and 14 January 1863, *Gazette,* 12 and 19 February 1863.
4. *Gazette,* 8 January and 12 February 1863.
5. *Ibid.,* 19 February and 9 April 1863.
6. *Ibid.,* 5 March 1863.
7. *Ibid.,* 19 February and 5 March 1863. *Preston,* pp. 161-162.
8. *Preston,* p. 161.
9. *Gazette,* 29 January 1863.
10. Samuel F. Atwill Diary, 27 February 1863, V.M.I. Archives.
11. *Gazette,* 25 March 1863. Minute Book of Rockbridge County, 2 February 1863.
12. *Gazette,* 2 April 1863. *Jones Diary,* 16 March and 1 April 1863.
13. Atwill Diary, 2 April 1863.
14. *Gazette,* 15 and 22 April 1863. *Preston,* p. 161.
15. Boney, pp. 191-194.
16. *O.R.,* Vol. 25, pp. 146-1056. *Preston,* p. 163. Atwill Diary, 5 May 1863.
17. *Preston,* p. 164. Atwill Diary, 9 May 1863.
18. *Preston,* p. 164.
19. *Ibid.,* pp. 164-165. Atwill Diary, 11 May 1863.
20. Atwill Diary, 14 May 1863 and copy of order in back of diary.
21. Wise, p. 270.
22. Atwill Diary, 20 May 1863.
23. *Gazette,* 20 May 1863.
24. Allan, p. 154.
25. *Ibid.*
26. *Gazette,* 21 May 1863.
27. *Preston,* pp. 166-167. *Gazette,* 19 August 1863.

28. *Preston*, pp. 167-168. For an account of refugees arriving in Rockbridge County during the war see: Mrs. Cornelia McDonald, *A Diary with Reminiscences of the War and Refugee Life in the Shenandoah Valley 1860-1865.* Edited and annotated by Hunter McDonald, Nashville, Tenn.: Cullom and Ghertner, 1934.
29. *1st and 2nd Rockbridge Artillery*, pp. 43-45 and 111-113. *52nd Virginia Infantry*, pp. 39-43. *Gazette*, 15 July 1863. *Preston*, p. 168. Allan, pp. 154-155.
30. Washington College Faculty Minutes, 3 July 1863.
31. *Military History of V.M.I.*, pp. 234-235. *Gazette*, 26 August 1863. *O.R.*, Vol. 29, pp. 49-50.
32. *O.R.*, Vol. 29, pp. 49-50. *Military History of V.M.I.*, pp. 235-239. *Gazette*, 26 August 1863. John G. Barrett and Robert K. Turner, Jr., eds. *Letters of A New Market Cadet: Beverly Stanard*, Chapel Hill: University of North Carolina Press, 1961, p. 9. Boney, p. 95.
33. *Gazette*, 26 August 1863. Wise, p. 272.
34. *Gazette*, 26 August, 6 September and 7 October 1863.
35. Unsigned letter from Engineering Dept. D.U.V. of 22 August 1863, James F. Jones letter of 4 February 1863, Major John A. Williams, Engineers, Richmond letter undated, Captain Theo. H. Tutwiler letters of 10 October 1862 and 4 August 1863, all to Presiding Justice, Rockbridge County, James K. Edmondson letter of 7 September 1863, Signed affidavit of free men detailed to Captain T. H. Tutwiler of 13 August 1863, all in Reid Family Papers, W&LU.
36. *O.R.*, Vol. 29, pp. 498-550. *Gazette*, 18 November 1863. *Preston*, pp. 168-169. H. McC. Davidson letter to "Dear Willie" 13 November 1863, Davidson Family Papers, W&LU. Jennings C. Wise, *Personal Memoir of Scott Shipp*, Lexington, Va.: 1915, p. 18.
37. *O.R.*, Vol. 29, pp. 498-550. *Gazette*, 18 November 1863. *Shipp*. pp. 18-21. William Couper, *One Hundred Years at V.M.I.*, 4 vols., Richmond: Garrett and Massie, 1939, pp. 217-227. *Stanard*, pp. 13-15. *Military History of V.M.I.*, pp. 254-263. *Preston*,pp. 168-171. H. McC. Davidson letter to "Dear Willie" 13 November 1863, Davidson Family Papers, W&LU.
38. *Preston*, pp. 168-171. McDonald, pp. 194-199. Minute Book of Rockbridge County, November and December 1863, January, March and April 1864.
39. *Gazette*, 24 February 1864.
40. Minutes of Lexington Town Council, December 1863 and 5 May 1864. Minute Book of Rockbridge County, December 1863.
41. *O.R.*, Vol. 29, pp. 919-973. *Military History of V.M.I.*, pp. 265-270. *Preston*, pp. 172-175. Allan, pp. 158-159. *Shipp*, pp. 28-29. *Stanard*, pp. 21-23. *Jones Diary*, 14 December 1863.
42. *O.R.*, Vol. 29, pp. 919-973. *Military History of V.M.I.* pp. 270-275. *Preston*, pp. 174-175. Allan, p. 159. *Stanard*, pp. 24-26.
43. *Preston*, p. 175. *Stanard*, pp. 28-35.
44. *Life Behind Confederate Lines in Virginia*, pp. 223-224. *Gazette*, 14 January and 17 February 1864. *Stanard*, pp. 31-33.

Chapter V

1. *Gazette*, 14 January 1864.
2. *Stanard*, pp. 31-33.
3. Charles W. Turner, ed. *My Dear Emma: War Letters of Colonel James K. Ed-*

mondson, 1861-1865, Vernona, Va.: McClure Press, 1978, pp. 128-129. Lee, p. 312. *Gazette,* 25 November 1863, 27 January, 10 and 17 February 1864. *52nd Virginia Infantry,* pp. 31-32. *The 1st and 2nd Rockbridge Artillery,* pp. 47 and 118.

4. Richard C. Todd, *Confederate Finance,* Athens, Ga.: 1954, pp. 11-12. *Preston,* pp. 177-178. Minute Book of Rockbridge County, May 1864. *Stanard,* pp. 26-27 and 40-42.
5. Thomas H. Williamson letter to General William H. Smith 25 March 1864 and John Haigh letter to "My Dear Mother" 24 March 1864, both in V.M.I. Archives.
6. *Military History of V.M.I.,* pp. 276-281. William Couper, *The Virginia Military Institute New Market Cadets,* Charlottesville: The Michie Co., 1933, pp. 373-375. *Stanard,* pp. 52-53. McDonald, pp. 200 and 220.
7. *Gazette,* 4 May 1864.
8. *Ibid.,* 11 May 1864.
9. *Ibid.*
10. *O.R.,* Vol. 37, pp. 77-91. *Military History of V.M.I.,* pp. 288-342. *Preston,* pp. 179-181. *Stanard,* pp. 61-70. McDonald, p. 201. *Shipp,* pp. 29-31. Wise, pp. 285-309. Smith,pp. 192-204.
11. Couper, *The V.M.I. New Market Cadets,* pp. 1-257. *The 1st and 2nd Rockbridge Artillery,* p. 77. *Military History of V.M.I.,* p. 337. *Stanard,* pp. 64-70.
12. *O.R.,* Vol. 37, pp. 744-74 and 747-748. *Military History of V.M.I.,* pp. 341-351.
13. *Gazette,* 11, 18 and 23 May 1864.
14. *Ibid.,* 23 May 1864.
15. *O.R.,* Vol. 37, pp. 93-160. Boatner, pp. 652-653. Marshall M. Brice, *Conquest of A Valley,* Charlottesville: University of Virginia Press, 1965, pp. 26-92 and 152-156. Robert J. Driver, Jr., *Staunton Artillery-McClanahan's Battery,* Lynchburg: H. E. Howard, Inc., 1988, pp. 90-98. Joseph A. Waddell, *Annals of Augusta County, Virginia,* 2nd edition, Harrisonburg: C. J. Carrier Co., 1972, pp. 488-490. *Preston,* p. 184. L. Bumgardner letter to "Dear Brother" 10 June 1864, W&LU Archives. *Gazette,* February 1899.
16. Minute Book of Rockbridge County, June 1864. Minute Book of the Lexington Town Council, 10 May 1864. *Gazette,* 18 May 1864.
17. *O.R.,* Vol. 37, pp. 93-160. John McCausland letter to Charles M. Blackford 30 January 1901, John Warwick Daniel Papers, University of Virginia. *14th Virginia Cavalry,* pp. 32-33.
18. *Preston,* pp. 184-187. *Jones Diary,* 8 June 1864. William H. Smith, Superintendent's Report to V.M.I. Board of Visitors 15 July 1864, V.M.I. Archives. L. Bumgardner Letter to "Dear Brother" 10 June 1864, W&LU Archives. Brice, pp. 93-108.
19. *O.R.,* Vol. 37, pp. 93-160. Waddell, pp. 490-493. Brice, pp. 114-115. *Life Behind Confederate Lines In Virginia,* pp. 224-228. *Military History of V.M.I.,* p. 353. *14th Virginia Cavalry,* p. 43. *Jones Diary,* 9 June 1864. James H. Stevenson, *Boots and Saddles: A History of the 1st Volunteer Cavalry of the War, Known as the 1st New York (Lincoln) Cavalry,* Harrisburg, Pa.: Patriot Printing Press, 1879, 10-11 June 1864. William H. Beach, *The First New York (Lincoln) Cavalry from April 19, 1861-July 7, 1865,* New York: The Lincoln Cavalry Association, 1902, pp. 369-370. Samuel A. Farrar, *The 22nd Pennsylvania Cavalry and the Ringgold Battalion 1861-1865,* Akron & Pittsburg: The 22nd Pennsylvania Cavalry Association, 1911, pp. 229-230.

20. *O.R.*, Vol. 37, pp. 93-160. Brice, pp. 114-115. *14th Virginia Cavalry*, pp. 33-34. Major Achilles J. Tynes letter to "My darling wife" 13 June 1864, Achilles J. Tynes Papers, Southern Historical Collection, University of North Carolina. *Military History of V.M.I.*, p. 353. *Jones Diary*, 9 June 1864. Brice, p. 115. Charles M. Keyes, *The Military History of the 123rd Regiment of Ohio Volunteer Infantry*, Sandusky: Register Steam Press, 1874, p. 65. David H. Strother, *A Virginia Yankee in the Civil War*, Cecil B. Eby, Jr., ed., Chapel Hill: University of North Carolina Press, 1961, p. 252.
21. Calvin Culton Memoir, 10 June 1864, Rockbridge Historical Society. *14th Virginia Cavalry*, pp. 34-35.
22. Mollie Patterson letter 9 July 1864, Patterson Family Papers, Rockbridge Historical Society. *Jones Diary*, 13 June 1864. Charles H. Lynch, *The Civil War Diary 1862-1865*, Hartford: Case, Lockwood & Brainard Co., 1915, 10 June 1864.
23. *Military History of V.M.I.*, pp. 354-355. *14th Virginia Cavalry*, pp. 34-35. *Preston*, p. 187. Wise, p. 310. *O.R*, Vol. 37, pp. 93-160.
24. *Military History of V.M.I.*, pp. 355-357. Wise, pp. 310-313. *14th Virginia Cavalry*, p. 35. General William H. Smith letter to General Richardson, 17 June 1864, Superintendent's Correspondence, V.M.I. Archives.
25. McDonald, pp. 203-204. *Preston*, p. 188. Russell Hastings' Reminiscences, 11 June 1864, 1900, Rutherford B. Hayes Library, Fremont, Ohio. Tynes letter, 13 June 1864. Lee, p. 344. White, p. 186. Fannie M. Lyle Wilson letter to "My dear Papa" 17 June 1864, *Rockbridge County News*, 1 November 1933. W. S. Agnor Memoir, *Gazette*, 5 July 1928.
26. *Military History of V.M.I.*, pp. 357-358. Wise, pp. 312-313. *Preston*, p. 188. Smith letter to General Richardson 17 June 1864.
27. *O.R.*, Vol. 37, pp. 93-160. Brice, p. 116. *14th Virginia Cavalry*, p. 35. *Preston*, pp. 188-189. Strother, pp. 252-253. Rose Pendleton Memoir, 18 June 1864, Elinor Gadsden Collection, W&LU Archives. Francis S. Reader (5th West Virginia Cavalry) Diary, 11 June 1864, W&LU Archives. Smith letter to General Richardson 17 June 1864. Allan, pp. 168-169.
28. Russell Hastings' Memoir, 11 June 1864. McDonald, pp. 204-205. Lynch, 11 June 1864. Dr. [Surgeon] John J. Booth Diary, 36th Ohio Infantry, 11 June 1864, Ohio State Historical Society. William C. Walker, *History of the 18th Connecticut Volunteers*, Norwich: 1885, 11 June 1864. McDonald, p. 221.
29. Rose Pendleton Memoir, 18 June 1864. Lee, pp. 344-345. McDonald, pp. 204-205.
30. Strother, p. 254. Agnor Memoir, *Gazette*, 5 July 1928.
31. *Preston*, pp. 189-192. Lee, pp. 346-349. Mrs. John R. Moore, *Memories of A Long Life In Virginia*, Staunton: The McClure Co., Inc., 1920. pp. 68-69.
32. *Preston*, p. 191. Strother, pp. 254-256. Reader Diary, 11-12 June 1864. Beach, pp. 370-371.
33. Smith letter to General Richardson 17 June 1864. Strother, p. 257. White, p. 186. McDonald, p. 206. Lee, p. 349.
34. Strother, p. 257. McDonald, pp. 207-208. Rose Pendleton Memoir 18 June 1864. Lynch Diary, 12 June 1864. Boney, pp. 207-208. Lee, p. 340. *Gazette*, 15 July 1864. Smith letter to General Richardson 17 June 1864. Margaret Letcher Showell, "Ex-Governor Letcher's Home: His Daughter Tells How it was Burn-

ed During the Civil War," *Southern Historical Society Papers,* Vol. XVIII (1890), pp. 393-397. Faculty Report to Board of Visitors, undated [circa June 1864], Washington College Faculty Records, W&LU. Booth Diary, 12-13 June 1864. Beach, p. 371. Sidney Martin, Union Soldier, letter 14 June 1864 and Benjamin F. Seibert, Signal Corps Detachment, Army of West Virginia, letter 13 June 1864, both in U.S. Army Military Research Collection, Carlisle, Pa. Charles R. Williams, ed., *Diary and Letters of Rutherford Birchard Hayes,* Vol. II, The Ohio State Archaeological and Historical Society, 1922, pp. 473-474.

35. White, pp. 186-187.
36. *Gazette,* 15 July 1864 and 5 July 1928. *Military History of V.M.I.,* pp. 370-371.
37. *Gazette,* 15 July 1864 and 1 November 1923. McDonald, p. 207. *Preston,* pp. 194-196. Lee, pp. 347-349. Rose Pendleton Memoir 18 June 1864. *14th Virginia Cavalry,* p. 35.
38. *Gazette,* 15 July 1864. Rose Pendleton Memoir 18 June 1864. Lee, p. 349. McDonald, p. 208.
39. Strother, pp. 257-258. William S. Lincoln, *Life with the 34th Massachusetts Infantry,* Worchester: Noyes Snow & Co., 1879, 13 June 1864. *Gazette,* 15 July 1864 and 5 July 1928. Rose Pendleton Memoir 18 June 1864. Reader Diary, 13 June 1864. *O.R.,* Vol. 37, p. 97.
40. Strother, p. 258. *Gazette,* 15 July 1864. *O.R.,* Vol. 37, p. 141. Farrar, pp. 241-242. Stevenson, 13 June 1864. Keyes, pp. 65-66. Smith letter to General Richardson 17 June 1864.
41. *O.R.,* Vol. 37, p. 97. Farrar, p. 242. *Gazette,* 15 July 1864. Strother, p. 257. White, p. 186. Smith letter to General Richardson 17 June 1864.
42. *O.R.,* Vol. 37, p. 97. Tynes letter 13 June 1864. *14th Virginia Cavalry,* pp. 35-37. Strother, pp. 257-259. Lee, p. 349. John D. Capron, "Virginia Iron Furnaces of the Confederacy," *Virginia Cavalcade Magazine,* 1966, p. 15.
43. *O.R.,* Vol. 37, p. 98. *Miltary History of V.M.I.,* p. 369. Reader Diary, 13 June 1864. Rose Pendleton Memoir 18 June 1864. Booth Diary, 14 June 1864. *Jones Diary,* 14 June 1864.
44. *Gazette,* 15 July 1864. *Preston,* p. 197.
45. McDonald, pp. 209-213. *Gazette,* 11 January 1927. Stevenson, 14 June 1864. Beach, pp. 372-373. Strother, pp. 259-261. Fannie M. Lyle Wilson letter 17 June 1864.
46. *Military History of V.M.I.,* pp. 372-381. Couper, *The V.M.I. New Market Cadets,* pp. 45-50. *Gazette,* 6 July 1864.
47. *Gazette,* 6 and 15 July 1864. Minute Book of Rockbridge County, July 1864. Minutes of the Lexington Town Council, 23 July 1864.
48. McDonald, p. 221. *52nd Virginia Infantry,* p. 63. Lee, p. 360-361.
49. *Gazette,* 19 July 2, 9 and 23, August 1864, 11 January and 6 May 1865. John B. Gibson and D. R. Reverly letter to Thomas H. Ellis, President of the J. R. and V. Co. 24 August 1864, Gibson Family Papers, Mark B. Williams, Petersburg, Va. Minute Book of Rockbridge County, November 1864.
50. McDonald, pp. 224-226. *Gazette,* 25 August 1864.
51. *Gazette,* 9 and 12 August, 6, 13, and 23 September, 12, 19, and 23 October and 2 November 1964. List of Exempts, Reserves and Substitutes from Rockbridge County, Rockbridge Historical Society.
52. *Gazette,* 16 August 1864. McDonald, p. 233.
53. Lee, p. 370-373. *Gazette,* 13 September and 19 October 1864.

54. McDonald, pp. 237-241. *Gazette*, 7 December 1864.
55. *Gazette*, 22 November 1864 and 4 January 1865.
56. *Ibid.*, 22 November and 21 December 1864. *14th Virginia Cavalry*, pp. 54-55. Lieutenant Colonel John A. Gibson Diary 1864-1865, Virginia Historical Society.
57. *Gazette*, 2 August and 7 December 1864, 11 January, 1 March and 13 April 1865. McDonald, pp. 240-250.
58. *Gazette*, 22 November 1864, 4 January and 22 February 1865. *52nd Virginia Infantry*, p. 72.
59. McDonald, pp. 245-254. Allan, p. 173. White, pp. 191-192. *Jones Diary*, 14 January 1865.

Chapter VI

1. White, pp. 194-195. Allan, pp. 174 and 176.
2. *Gazette*, 11 January 1865. *14th Virginia Cavalry*, p. 60.
3. *Gazette*, 18 January 1865. Minutes of the Lexington Town Council, 7 January 1865. Minute Book of Rockbridge County, December 1864.
4. *Gazette*, 4 January and 14 February 1865. *Jones Diary*, 1 January and 7 February 1865.
5. *Gazette*, 1 March 1865. *1st and 2nd Rockbridge Artillery*, pp. 122-123.
6. *Gazette*, 22 February 1865. McDonald, pp. 254-255.
7. *Gazette*, 1 and 31 March 1865.
8. *Ibid.*, 31 March 1865.
9. *Ibid. 14th Virginia Cavalry*, p. 60. McDonald, pp. 252-253. List of Exempts, Reserves and Substitutes from Rockbridge County, Rockbridge Historical Society.
10. *Gazette*, 13 April 1865. *1st and 2nd Rockbridge Artillery*, pp. 137-138, 53-56 and 123-125. White, p. 197. Allan, pp. 185-188. Preston, p. 207.
11. *14th Virginia Cavalry*, pp. 62-66.
12. *Gazette*, 13 and 20 April 1865.
13. *Ibid.*, 20 April and 4 May 1865. McDonald, pp. 262-264.
14. John M. Brooke, *General Lee's Church: The History of the Protestant Episcopal Church in Lexington, Virginia 1840-1975.* Lexington: *The News-Gazette*, 1984, pp. 18-19. McDonald, p. 264.
15. Boney, pp. 217-233.
16. Minutes of the Lexington Town Council, 31 May, 28 June, 23 July and 29 November 1865. *Gazette*, 2 August 1865. McDonald, pp. 263-264.
17. *Gazette*, 2 August, 20 September and 22 November 1865. *Edmondson*, 135-137.
18. *Gazette*, 13 May and 2 August 1865.
19. Robert E. Lee, Jr., *Recollections and Letters of General Robert E. Lee*, New York: Doubleday, Page & Co., 1905, pp. 179-188. *Gazette*, 20 September and 11 October 1865. Board of Trustees, Washington College, Resolution 31 August 1865 and John W. Brockenbrough letter to "General" 31 August 1865, both in W&LU Archives. Allan, pp. 199-201. McDonald, pp. 268-269.
20. *Gazette*, 11 October 1865. Minutes of the Lexington Town Council, 15 December 1865, 6 January and 12 March 1866.
21. *Gazette*, 29 November and 6 December 1865, 10 and 17 January, 14 February 28 March and 18 April 1866.
22. *Ibid.*, 31 January, 16 and 23 May 1866.

BIBLIOGRAPHY

Manuscripts

Rutherford B. Hayes Library, Freemont, Ohio.
    Russell Hastings' Reminiscences
Indiana State Library.
    J. H. Davidson Collection
Lexington City Hall.
    Minutes of the Lexington Town Council 1860-1867
University of North Carolina.
    Southern Historical Collection
        Major Archilles J. Tynes Papers
Ohio State Historical Society.
    Dr. John J. Booth Diary, 36th Ohio Volunteers
Rockbridge County Courthouse.
    Applications for Commutation of Wounds Under the Act of 1884
    Civil War Papers
    Minute Book of Rockbridge County 1860-1867
    Order Book of Rockbridge County 1860-1867
    Records of the Sons of Confederate Veterans Camp, Lexington
    Roster of Companies of Rockbridge County Men in the War Between the States
Rockbridge County Historical Society.
    Cemetery Listings
    Civil War Papers
    Confederate Veterans Papers
    Calvin Culton Memoir
    Davidson Family Papers
    Houston Family Papers
    Hunter's Raid (File)
    John Letcher Papers
    List of Exempts, Reserves and Substitutes from Rockbridge County
    List of Participants in the Civil War from Rockbridge County
    John McCausland Letters
    Miley Collection (Photographs)
    Muster Rolls First Battalion, Eighth Regiment, Virginia Militia 1841-1862
    Muster rolls of various companies from Rockbridge County
    Patterson Family Papers
    Records of Lee-Jackson Camp, Confederate Veterans, Lexington
    Records of the Monmouth Cloth Factory

United Daughters of the Confederacy Records

United States Army Military Research Center, Carlisle, Pennsylvania.

Sidney Martin Letter

Benjamin F. Seebert Letter

Virginia Historical Society.

John A. Gibson Diary 1864-1865

Virginia Military Institute.

Alumni and Faculty Records

Samuel F. Atwill Diary

Board of Visitors Records 1860-1868

A. C. S. Gatewood Letters

Journal of Colonel James T. L. Preston

T. A. Stevenson Letters

Superintendents Correspondence 1860-1868

Waddell Family Papers

James J. White Letters

Mary L. R. White Letters

Virginia State Library.

Compiled Roster of Regiments, Battalions and Companies from Virginia

Joseph Bidgood Papers, Unit Record Department, Adjutant General's Office, Confederate Military Records

Muster Rolls of various companies from Rockbridge County

Pension Applications, Acts of 1888, 1900 and 1902 from Rockbridge County

University of Virginia.

John Warwick Daniels Papers

McGuffin Family Papers

A. Patterson Papers

Washington and Lee University.

Alexander T. Barclay Letters

Davidson Family Papers

George West Diehl Papers

Elinor Gadsden Collection

Letcher Family Papers

McDowell Papers

Miley Collection (Photographs)

Pendleton Family Papers

Francis S. Reeder Diary

Reid Family Papers

Templeton Family Papers

Washington College Board of Visitor Records 1860-1868

Washington College Faculty Correspondence 1860-1868

Washington College Faculty Minutes 1860-1868

Mark William Collection, Petersburg, Virginia.
Gibson Family Papers
Wisconsin Historical Society.
Davidson Family Papers

Public Documents
Virginia

Buena Vista Census: 1900
Rockbridge County Censuses: 1860, 1870 and 1910.

Periodicals

All The Year Round. Charles Dickens, editor. "William W. Greenlee Diary 1861." London, England: 17 May 1862.

Civil War History, Bruce B. Greenwalt, editor. "Life Behind Confederate Lines in Virginia: The Correspondence of James Dorman Davidson." Volume 16, September, 1970.

Confederate Veteran, 1893-1932. 40 volumes.

Southern Historical Society Papers. Richmond, 1876-1953. 52 volumes.

The Ironworker. Volume XXVIII, No. 2, Lynchburg, Spring 1964.

Virginia Cavalcade Magazine. John D. Capon, "Virginia Iron Furnaces of the Confederacy." 1966.

The Virginia Magazine of History and Biography, Wiliam G. Bean, editor. "A House Divided: The Civil War Letters of a Virginia Family." Volume 59, October, 1959.

Newspapers

Buena Vista *Advocate*
Glasgow *Herald*
Goshen *Blade*
Lexington *Gazette and General Advertiser*
Lexington *Gazette*
Richmond *Sentinel*
Rockbridge County News
Rockbridge Enterprise
Staunton *Vindicator*
Staunton *Spectator*
The Rockbridge Citizen
Valley Star

Published Works

Albaugh, William A. III and Simmons, Edward N. *Confederate Arms.* New York: Bonanza Books, 1957.

Allan, Elizabeth Randolph Preston. *The Life and Letters of Margaret Junkin Preston.* New York: Houghton-Mifflin, 1903.

________. *A March Past: Reminiscences of Elizabeth Randolph Preston Allan.* Janet Allan Bryant, ed. Richmond: The Dietz Press, 1938.

Armstrong, Richard L. *Ambush at Williamsville.* Staunton, Virginia: Minuteman Press, 1986.

Barrett, John G. and Robert K. Turner, Jr., eds. *Letters of A New Market Cadet: Beverley Stanard.* Chapel Hill: University of North Carolina Press, 1961.

Beach, William H. *The First New York (Lincoln) Cavalry from April 19, 1861-July 7, 1865.* New York: The Lincoln Cavalry Association, 1902.

Bell, Robert T. *11th Virginia Infantry.* Lynchburg: H. E. Howard, Inc., 1985.

Boatner, Mark M., III. *The Civil War Dictionary.* New York: David McKay Co., Inc., 1966.

Boley, Henry. *Lexington in Old Virginia.* Richmond: Garrett & Massie Publishers, 1935.

Boney, F. N. *John Letcher of Virginia.* University, Alabama: University of Alabama Press, 1966.

Brice, Marshall M. *Conquest of a Valley.* Charlottesville: University of Virginia Press, 1965.

Brooke, John M. *General Lee's Church: The History of the Protestant Episcopal Church in Lexington, Virginia 1840-1975.* Lexington: *The News Gazette,* 1985.

Chambers, Lenoir. *Stonewall Jackson and the Virginia Military Institute: The Lexington Years.* Lexington: Historic Lexington Foundation, 1982.

Couper, William. *One Hundred Years at V.M.I.* 4 volumes. Richmond: Garrett & Massie Publishers, 1939.

________. *The Virginia Military Institute New Market Cadets.* Charlottesville: The Michie Co., 1933.

Cupp, Albert M. *A History of Methodism in Rockbridge County, Virginia.* n.p., n.d.

Davis, Thomas W., ed. *A Crowd of Honorable Youths.* Lexington: Virginia Military Institute Sesquicentennial Committee, 1988.

Dew, Charles B. *Ironmaker to the Confederacy, Joseph R. Anderson and the Tredegar Ironworks.* New Haven: Yale University Press, 1966.

Driver, Robert J., Jr. *52nd Virginia Infantry.* Lynchburg: H. E. Howard, Inc. 1986.

________. *1st and 2nd Rockbridge Artillery.* Lynchburg: H. E. Howard, Inc., 1987.

________. *14th Virginia Cavalry.* Lynchburg: H. E. Howard, Inc. 1988.

________. *Staunton Artillery-McClanahan's Battery.* Lynchburg: H. E. Howard, Inc., 1988.

DuPont, Henry A. *The Campaign of 1864 in the Valley of Virginia and the Expedition to Lynchburg.* New York: National Americana Society, 1925.

Farrar, Samuel A. *The 22nd Pennsylvania Cavalry and the Ringgold Battalion 1861-1865.* Akron & Pittsburg: The 22nd Pennsylvania Cavalry Association, 1911.

Hayes, Rutherford B. *Diary and Letters of Rutherford B. Hayes.* Charles R. Edwards, ed. Volume 2. The Ohio State Archaeological and Historical Society, 1922.

Hoar, Jay S. *The South's Last Boys in Gray.* Bowling Green, Ohio: Bowling Green State University Press, 1987.

Johnson, J. E., ed. *Echos of the Virginia Military Institute.* Lexington: n.p., 1937.

Johnson, Robert U. and C. C. Buel, editors. *Battles and Leaders of the Civil War.* 4 volumes. New York: Thomas Yoseloff, 1956.

Jones, J. William. *Christ in the Camp.* Richmond: B. F. Johnson & Co., 1887.

Junkin, Rev. E. D. *A History of the Church and Congregation of New Providence, Lexington Presbytery. Virginia:* n.p., 1871.

Junkin, George. *David X. Junkin, D. D., LL. D., A Historical Biography.* Philadelphia: 1871.

Keyes, Charles M. *The Military History of the 123rd Regiment of Ohio Volunteer Infantry.* Sandusky: Register Steam Press, 1874.

Lee, Robert E., Jr. *Recollections and Letters of General Robert E. Lee.* New York: Doubleday, Page & Co., 1905.

Lee, Susan P. *Memoirs of William Nelson Pendleton, D.D.* Philadelphia: J. B. Lippencott Co., 1893.

Lincoln, William S. *Life with the 34th Massachusetts Infantry.* Worchester: Noyes Snow & Co., 1879.

Lynch, Charles H. *The Civil War Diary 1862-1865.* Hartford: Case, Lockwood & Brainard Co., 1915.

McDonald, Mrs. Cornelia. *A Diary with the Reminiscences of the War and Refugee Life in the Shenandoah Valley.* Nashville: Cullom and Ghertner Co., 1934.

Moore, Mrs. John H. *Memories of a Long Life in Virginia.* Staunton: The McClure Co., Inc., 1920.

Morton, Oren F. *A History of Rockbridge County, Virginia.* Baltimore: Regional Publication Co., 1971.

Paxton, Alexander S. *Memory Days.* New York and Washington: The Neal Publishing Co., 1908.

Robertson, James I., Jr. *4th Virginia Infantry.* Lynchburg, H. E. Howard, Inc., 1982.

Smith, Francis H. *History of the Virginia Military Institute, Its Building and Rebuilding.* Lynchburg: J. P. Bell Co., 1912.

Stevenson, James H. *Boots and Saddles: A History of the 1st Volunteer Cavalry of the War, Known as the 1st New York (Lincoln) Cavalry.* Harrisburg, Pa.: Patriot Printing Press, 1879.

Strother, David H. *A Virginia Yankee In the Civil War, The Diaries of David Hunter Strother.* Cecil B. Eby, Jr., ed. Chapel Hill: University of North Carolina Press, 1961.

Taylor, George Baxter. *Virginia Baptist Ministers.* Lynchburg: J. P. Bell Co., 1915.

Todd, Richard C. *Confederate Finance.* Athens, Ga.: 1954.

Tompkins, Edmund Pendleton. *Rockbridge County, Virginia: An Informal History.* Richmond: Whittet & Shepperson, 1952.

Turner, Charles W., ed. *Captain Greenlee Davidson, C.S.A. Diary and Letters 1851-1863.* Verona, Va.: McClure Press, 1975.

_______ ed. *My Dear Emma: War Letters of Colonel James K. Edmondson, 1861-1865.* Verona, Va.: McClure Press, 1978.

_______ ed. *The Diary of Henry Boswell Jones of Brownsburg, 1842-1871.* Verona, Va.: McClure Press, 1979.

_______ ed. *Lieutenant John Newton Lyle: The Career of the Liberty Hall Volunteers.* Verona, Va.: McClure Press, 1986.

_______. *Virginia's Green Revolution.* Waynesboro, Va.: The Humphries Press, Inc., 1986.

_______. ed. *Old Zeus: Life and Letters of James J. White.* Verona, Va.: McClure Printing Co., 1983.

U.S. War Department. *The War of the Rebellion: A Compilation of the Official Records of the Union and Confederate Armies.* 128 volumes. Washington, D.C.: Government Printing Office, 1880-1901.

Virginia Military Institute. *Virginia Military Institute Register of Former Cadets.* Lexington: 1957.

Waddell, Joseph A. *Annals of Augusta County, Virginia.* Staunton: C. R. Caldwell, 1902.

Walker, William C. *History of the 18th Connecticut Volunteers.* Norwich: 1885.

Wallace, Lee A., Jr. *A Guide to Virginia Military Organizations, 1861-1865.* Lynchburg: H. E. Howard, Inc., 1986.

_______. *5th Virginia Infantry.* Lynchburg: H. E. Howard, Inc., 1988.

Washington and Lee University Historical Papers. No. 6, 1904. Lynchburg: J. P. Bell Co., 1904.

White, H. M., ed. *Reverend William S. White, D.D., and His Times.* Richmond: Presbyterian Committee of Publication, 1891.

White, William S. *Sketches of the Life of Captain Hugh A. White, of the Stonewall Brigade.* Columbia, S.C.: 1864.

Wise, Jennings C. *Personal Memoir of Scott Shipp.* Lexington: n.p. 1915.

_______. *The Military History of the Virginia Military Institute from 1839-1865.* Lynchburg: J. P. Bell Co., 1915.

Wise, John S. *The End of An Era.* Boston & New York: Houghton-Mifflin Co., 1899.

Wood, James H. *The War, Stonewall Jackson: his Campaigps and Battles, the Regiment as I Saw Them.* Cumberland, Md.: The Eddy Press, 1916.

INDEX

www.ingramcontent.com/pod-product-compliance
Lightning Source LLC
LaVergne TN
LVHW050643100826
845148LV00011B/1965
9780788430282